Science Breakthroughs to Advance Food and Agricultural Research by 2030

Committee on Science Breakthroughs 2030:
A Strategy for Food and Agricultural Research

Board on Agriculture and Natural Resources
Board on Atmospheric Sciences and Climate
Board on Life Sciences
Water Science and Technology Board
Division on Earth and Life Studies

Food and Nutrition Board
Health and Medicine Division

Board on Environmental Change and Society
Division of Behavioral and Social Sciences and Education

A Consensus Study Report of

The National Academies of
SCIENCES · ENGINEERING · MEDICINE

THE NATIONAL ACADEMIES PRESS
Washington, DC
www.nap.edu

THE NATIONAL ACADEMIES PRESS 500 Fifth Street, NW Washington, DC 20001

This study was supported by Supporters of Agricultural Research Foundation (#201701), the National Institute of Food and Agriculture of the U.S. Department of Agriculture (#2017-38886-26911), the National Science Foundation (IOS-1747820), and the U.S. Department of Energy (DE-FOA-0001820), with additional funding from the Foundation for Food and Agriculture Research. Any opinions, findings, conclusions, or recommendations expressed in this publication do not necessarily reflect the views of any organization or agency that provided support for the project.

International Standard Book Number-13: 978-0-309-47392-7
International Standard Book Number-10: 0-309-47392-6
Digital Object Identifier: https://doi.org/10.17226/25059
Library of Congress Control Number: 2018961240

Additional copies of this publication are available from the National Academies Press, 500 Fifth Street, NW, Keck 360, Washington, DC 20001; (800) 624-6242 or (202) 334-3313; http://www.nap.edu.

Printed in the United States of America

Suggested Citation: National Academies of Sciences, Engineering, and Medicine. 2019. *Science Breakthroughs to Advance Food and Agricultural Research by 2030*. Washington, DC: The National Academies Press. doi: https://doi.org/10.17226/25059.

The National Academies of

SCIENCES · ENGINEERING · MEDICINE

COMMITTEE ON SCIENCE BREAKTHROUGHS 2030: A STRATEGY FOR FOOD AND AGRICULTURAL RESEARCH

Co-Chairs

JOHN D. FLOROS, New Mexico State University, Las Cruces
SUSAN R. WESSLER (NAS), University of California, Riverside

Members

DAVID B. ALLISON (NAM), Indiana University School of Public Health–Bloomington
CORRIE C. BROWN, University of Georgia, Athens
LISA M. GODDARD, Columbia University, New York
MARY LOU GUERINOT (NAS), Dartmouth College, Hanover, NH
JANET K. JANSSON, Pacific Northwest National Laboratory, Richland, WA
LEE-ANN JAYKUS, North Carolina State University, Raleigh
HELEN H. JENSEN, Iowa State University, Ames
RAJIV KHOSLA, Colorado State University, Fort Collins
ROBIN LOUGEE, IBM Research, Yorktown Heights, NY
GREGORY V. LOWRY, Carnegie Mellon University, Pittsburgh, PA
ALISON L. VAN EENENNAAM, University of California, Davis

Staff

PEGGY TSAI YIH, Study Director, Board on Agriculture and Natural Resources
MARIA ORIA, Senior Program Officer, Food and Nutrition Board
AMANDA PURCELL, Program Officer, Board on Atmospheric Sciences and Climate
KEEGAN SAWYER, Senior Program Officer, Board on Life Sciences
ROBIN SCHOEN, Director, Board on Agriculture and Natural Resources
TOBY WARDEN, Director, Board on Environmental Change and Society
YASMIN ROMITTI, Research Associate, Board on Atmospheric Sciences and Climate
JENNA BRISCOE, Research Assistant, Board on Agriculture and Natural Resources
ERIN MARKOVICH, Research Assistant, Board on Atmospheric Sciences and Climate

BOARD ON AGRICULTURE AND NATURAL RESOURCES

Chair

CHARLES W. RICE, Kansas State University, Manhattan

Members

SHANE C. BURGESS, University of Arizona, Tucson
SUSAN CAPALBO, Oregon State University, Corvallis
GAIL CZARNECKI-MAULDEN, Nestlé Purina PetCare, St. Louis, MO
GEBISA EJETA, Purdue University, West Lafayette, IN
JAMES S. FAMIGLIETTI, California Institute of Technology, Pasadena
FRED GOULD (NAS), North Carolina State University, Raleigh
DOUGLAS B. JACKSON-SMITH, The Ohio State University, Wooster
JAMES W. JONES (NAE), University of Florida, Gainesville
STEPHEN S. KELLEY, North Carolina State University, Raleigh
JAN E. LEACH, Colorado State University, Fort Collins
JILL J. McCLUSKEY, Washington State University, Richland
KAREN I. PLAUT, Purdue University, West Lafayette, IN
JIM E. RIVIERE (NAM), Kansas State University, Manhattan

Staff

ROBIN A. SCHOEN, Director
CAMILLA YANDOC ABLES, Senior Program Officer
JENNA BRISCOE, Research Assistant
KARA N. LANEY, Senior Program Officer
PEGGY TSAI YIH, Senior Program Officer

Acknowledgments

This Consensus Study Report was reviewed in draft form by individuals chosen for their diverse perspectives and technical expertise. The purpose of this independent review is to provide candid and critical comments that will assist the National Academies of Sciences, Engineering, and Medicine in making each published report as sound as possible and to ensure that it meets the institutional standards for quality, objectivity, evidence, and responsiveness to the study charge. The review comments and draft manuscript remain confidential to protect the integrity of the deliberative process.

We thank the following individuals for their review of this report:

Molly D. Anderson, Middlebury College
Brian T. Cunningham, University of Illinois
Charles J. Czuprynski, University of Wisconsin–Madison
Jeffery L. Dangl (NAS), University of North Carolina at Chapel Hill
Mary K. Firestone (NAS), University of California, Berkeley
Fred Gould (NAS), North Carolina State University
Gerrit Hoogenboom, University of Florida
Chandra Krintz, University of California, Santa Barbara
Marc B. Parlange (NAE), Monash University
Joseph D. Puglisi (NAS), Stanford University School of Medicine
Donald W. Schaffner, Rutgers University
Daniel Stokols, University of California, Irvine
Harold M. van Es, Cornell University
David Zilberman, University of California, Berkeley

Although the reviewers listed above provided many constructive comments and suggestions, they were not asked to endorse the conclusions or recommendations of this report nor did they see the final draft before its release. The review of this report was overseen by Dr. Michael T. Clegg, University of California, Irvine, and Dr. Norman R. Scott, Cornell University. They were responsible for making certain that an independent examination of this report was carried out in accordance with the standards of the National Academies and that all review comments were carefully considered. Responsibility for the final content rests entirely with the authoring committee and the National Academies.

Contents

APPENDIXES

Summary

1. INTRODUCTION

For nearly a century, scientific advances have fueled progress in U.S. agriculture to enable American producers to deliver safe and abundant food domestically and provide a trade surplus in bulk and high-value agricultural commodities and foods. Today, the U.S. food and agricultural enterprise faces formidable challenges that will test its long-term sustainability, competitiveness, and resilience. On its current path, future productivity in the U.S. agricultural system is likely to come with trade-offs. The success of agriculture is tied to natural systems, and these systems are showing signs of stress, even more so with the change in climate. Water scarcity, increased weather variability, floods, and droughts are examples of stresses on food and agricultural production. More than one-third of the food produced is unconsumed, an unacceptable loss of food and nutrients at a time of heightened global food demand. Increased food animal production to meet greater demand will generate more greenhouse gas emissions and excess animal waste. The U.S. food supply is generally secure, but is not immune to the costly and deadly shocks of continuing outbreaks of foodborne illness or to the constant threat of pests and pathogens to crops, livestock, and poultry. U.S. farmers and producers are at the front lines and will need more tools to manage the pressures they face.

In the coming decade, stresses on the U.S. food and agricultural enterprise are unlikely to be resolved by farmers, the market, input suppliers, or by current public- and private-sector research efforts if business as usual prevails. Approaches focused mainly on making incremental fixes to problems that arise from complex influences—some biological and physical, some man-made—are resistant to simple solutions. The food system is vast, complex, and interconnected. The "wicked" problems—intractable problems with many interdependent factors that make them difficult to define or solve—will require a radically different approach to understand and uncover solutions that can only be found when explored beyond the traditional boundaries of food and agricultural disciplines. Broader perspectives are needed to provide a better view for optimizing the food and agricultural system. Acquiring this perspective means reframing problems and employing emerging tools to identify and address key points of intervention in the system.

This report identifies innovative, emerging scientific advances for making the U.S. food and agricultural system more efficient, resilient, and sustainable. An ad hoc study committee appointed by the National Academies of Sciences, Engineering, and Medicine was guided by a Statement of Task[1] to explore the availability of relatively new scientific developments across all disciplines that could accelerate progress toward those goals. The committee identified the most promising scientific breakthroughs that could have the greatest positive impact on food and agriculture, and that are possible to achieve in the next decade (by 2030). The opportunities summarized in this report highlight novel approaches for food and agricultural sciences in innovating for the future.

2. RESEARCH STRATEGY FOR 2030

2.1 Major Goals and Key Research Challenges

Over the course of its study, the committee held discussions with members of the scientific community to identify the most challenging issues facing food and agriculture and the best research opportunities to address them. In the next decade, the major goals for food and agricultural research include (1) improving the efficiency of food and agricultural systems, (2) increasing the sustainability of agriculture, and (3) increasing the resiliency of agricultural systems to adapt to rapid changes and extreme conditions. These goals derive from the common nature of key research challenges identified by food and agricultural scientists, which include the following:

- Increasing nutrient use efficiency in crop production systems,
- Reducing soil loss and degradation,
- Mobilizing genetic diversity for crop improvement,
- Optimizing water use in agriculture,

[1]The Statement of Task is provided in Chapter 1, Box 1-2.

- Improving food animal genetics,
- Developing precision livestock production systems,
- Early and rapid detection and prevention of plant and animal diseases,
- Early and rapid detection of foodborne pathogens, and
- Reducing food loss and waste throughout the supply chain.

2.2 Convergence

In the past, it has been more common to examine problems in a defined space or discipline for reasons related to practicality and greater ease of management, and that approach has been effective at addressing distinct issues that require specific knowledge in a domain. The urgent progress needed today to address the most challenging problems requires leveraging capabilities across the scientific and technological enterprise in a convergent research approach. The 2014 National Research Council report *Convergence: Facilitating Transdisciplinary Integration of Life Sciences, Physical Sciences, Engineering, and Beyond* (see p. 1) describes convergence as

> an approach to problem solving that cuts across disciplinary boundaries [and] integrates knowledge, tools, and ways of thinking from life and health sciences, physical, mathematical, and computational sciences, engineering disciplines, and beyond to form a comprehensive synthetic framework for tackling scientific and societal challenges that exist at the interfaces of multiple fields.

This means that merging diverse expertise areas stimulates innovation in both basic science discoveries and translational applications. Food and agricultural research needs to be broadened to harness advances in data science, materials science, and information technology. Furthermore, integrating the social sciences (such as behavioral and economics sciences) to correctly frame problems and their solution space is essential, as the food and agricultural system is as much a human system as a biophysical one.

2.3 Science Breakthroughs and Recommendations

The committee identified five breakthrough opportunities that could dramatically increase the capabilities of food and agricultural science. The recommendations that follow will require a shift in how the research community approaches its work, and initiatives for each of the breakthroughs will require robust support.

Transdisciplinary Research and Systems Approach

Breakthrough 1: A systems approach to understand the nature of interactions among the different elements of the food and agricultural system can be leveraged to increase overall system efficiency, resilience, and sustainability. Progress in meeting major goals can occur only when the scientific community begins to more methodically integrate science, technology, human behavior, economics, and policy into biophysical and empirical models. For example, there is the need to integrate the rate and determinants of adopting new technologies, practices, products, and processing innovations into food and agricultural system models. This approach is required to properly quantify the shifts in resource use, market effects, and response, and to determine benefits that are achievable from scientific and technological breakthroughs. Consideration of these system interactions is critical for finding holistic solutions to the food and agricultural challenges that threaten our security and competitiveness.

Recommendation 1: Transdisciplinary science and systems approaches should be prioritized to solve agriculture's most vexing problems. Solving the most challenging problems in agriculture will require convergence and systems thinking to address the issues; in the absence of both, enduring solutions may not be achievable. Transdisciplinary problem-based collaboration (team science) will need to be facilitated because for some, it is difficult to professionally gravitate to scientific fields outside of one's expertise. Such transitions will require learning to work in transdisciplinary teams. Enticing and enabling researchers from disparate disciplines to work effectively together on food and agricultural issues will require incentives in support of the collaboration. The use of convergent approaches will also facilitate new collaborations that may not have occurred when approached by researchers operating in disciplines in separate silos. Transdisciplinary problem-based collaborations will enable engagement of a new or diverse set of stakeholders and partners and benefit the food and agriculture sector. Leadership is key to making team science successful, as scientific directors need a unique set of skills that includes openness to different perspectives, the ability to conceptualize the big picture, and perhaps most importantly, a talent for uniting people around a common mission. These qualities are not always natural for scientists, so providing professional development opportunities to foster leadership in the transdisciplinary model is critical.

There are many examples of programs that already require transdisciplinary work: for example, grants provided by the National Science Foundation's (NSF's) Innovations at the Nexus of Food, Energy and Water Systems (INFEWS) and the request for proposals outlined in the 2018 Sustainable Agricultural Systems competitive grants program administered through the U.S. Department of Agriculture's (USDA's) Agricultural and

Food Research Initiative. The NSF INFEWS and most recent USDA grants on Sustainable Agricultural Systems have relatively larger budgets that can support convergent team science. However, many of the standard grants requiring "transdisciplinary" approaches do not provide enough funding to support team science so the incentives for transdisciplinary science are still lacking. For convergence to truly be productive, financial incentives are needed to encourage grant applicants to step outside their comfort zones and to establish deep connections among subject matter experts from a variety of arenas.

Sensing Technologies

Breakthrough 2: The development and validation of precise, accurate, field-deployable sensors and biosensors will enable rapid detection and monitoring capabilities across various food and agricultural disciplines. Historically, sensors and sensing technology have been used in food and agriculture to provide point measurements for certain characteristics of interest (e.g., temperature), but the ability to continuously monitor several characteristics at once is the key to understanding both what and how it is happening in the target system. Scientific and technological advances in materials science, microelectronics, and nanotechnology are poised to enable the creation of novel nano- and biosensors to continuously monitor conditions of environmental stimuli and biotic and abiotic stresses. The next generation of sensors may also revolutionize the ability to detect disease prior to the onset of symptoms in plants and animals, to identify human pathogens before they enter the food distribution chain, and to monitor and make decisions in near real time.

Recommendation 2: Create initiatives to more effectively employ existing sensing technologies and to develop new sensing technologies across all areas of food and agriculture. These initiatives would lead to transdisciplinary research, development, and application across the food system. The attributes of the sensor (e.g., shape, size, material, in situ or in planta, mobile, wired or wireless, biodegradable) would depend on the purpose, application, duration, and location of the sensors. For example, in situ soil and crop sensors may provide continuous data feed and may alert the farmer when moisture content in soil and turgor pressure in plants fall below a critical level to initiate site-specific irrigation to a group of plants, eliminating the need to irrigate the entire field. Likewise, in planta sensors may quantify biochemical changes in plants caused by an insect pest or a pathogen, alerting and enabling the producer to plan and deploy immediate site-specific control strategies before infestation occurs or damage is visible. Biosensors for food products could indicate product adulteration or spoilage and could alert distributors and consumers to take necessary action.

Data Science and Agri-Food Informatics

Breakthrough 3: The application and integration of data science, software tools, and systems models will enable advanced analytics for managing the food and agricultural system. The food and agricultural system collects an enormous amount of data, but has not had the right tools to use it effectively. Data generated in research laboratories and in the field have been maintained in an unconnected manner, preventing the ability to generate insights from their integration. Advances and applications of data science and analytics have been highlighted as an important breakthrough opportunity to elevate food and agricultural research and the application of knowledge. The ability to more quickly collect, analyze, store, share, and integrate highly heterogeneous datasets will create opportunities to vastly improve our understanding of the complex problems and, ultimately, to the widespread use of near-real-time, data-driven management approaches.

Recommendation 3: Establish initiatives to nurture the emerging area of agri-food informatics and to facilitate the adoption and development of information technology, data science, and artificial intelligence in food and agricultural research. Maximizing the knowledge and utility that can be gained from large research datasets requires strategic efforts to provide better data access, data harmonization, and data analytics in food and agricultural systems. The challenges of handling massive datasets that are highly heterogeneous across space and time need to be addressed. Data standards need to be established and the vast array of data needs to be more findable, interoperable, and reuseable. There is a need to increase data processing speeds, develop methods to quickly assess data veracity, and provide support for the development and dissemination of agri-food informatics capabilities, including tools for modeling real-time applications in dynamically changing conditions.

Blockchain and artificial intelligence, including machine-learning algorithms, are promising technologies for the unique needs of the food and agricultural system that have yet to be fully developed. Development of advanced analytic approaches, such as machine-learning algorithms for automated rapid phenotyping, will require better platforms for studying how various components in the food system interact. Application of these approaches will require investment in infrastructure to house massive numbers of records, and a means by which those records can be integrated and effectively used for decision-making purposes. A convergence of expertise from many disciplines will be needed to realize the potential of these opportunities.

Genomics and Precision Breeding

Breakthrough 4: The ability to carry out routine gene editing of agriculturally important organisms will allow for precise and rapid improvement of traits important for productivity and quality. Gene editing—aided by recent advances in genomics, transcriptomics, proteomics, and metabolomics—is poised to accelerate breeding to generate traits in plants, microbes, and animals that improve efficiency, resilience, and sustainability. Comparing hundreds of genotypes using omics technologies can speed the selection of alleles to enhance productivity, disease or drought resistance, nutritional value, and palatability. For instance, the tomato metabolome was effectively modified for enhanced taste, nutritional value, and disease resistance, and the swine genome was effectively targeted with the successful introduction of resistance to porcine reproductive and respiratory syndrome virus. This capability opens the door to domesticating new crops and soil microbes, developing disease-resistant livestock, controlling organisms' response to stress, and mining biodiversity for useful genes.

Recommendation 4: Establish initiatives to exploit the use of genomics and precision breeding to genetically improve traits of agriculturally important organisms. Genetic improvement programs in crops and animals are an essential component of agricultural sustainability. With the advent of gene-editing technologies, targeted genetic improvements can be applied to plant and animal improvement in a way that traditional methods of modification are unable to achieve. There are opportunities to accelerate genetic improvement by incorporating genomic information, advanced breeding technologies, and precision breeding methods into conventional breeding and selection programs. Encouraging the acceptance and adoption of some of these breakthrough technologies requires insights gained from social science and related education and communication efforts with producers and the public. Gene editing could be used to both expand allelic variation introduced from wild relatives into crops and remove undesirable linked traits, thereby increasing the value of genetic variation available in breeding programs. Similarly, incorporating essential micronutrients or other quality-related traits in crops through gene-editing tools offers an opportunity to increase food quality and shelf life, enhance nutrition, and decrease food loss and food waste. These technologies are similarly applicable to food animals, and possible targets of genetic improvements include enhanced fertility, removal of allergens, improved feed conversion, disease resistance, and animal welfare.

Microbiome

Breakthrough 5: Understand the relevance of the microbiome to agriculture and harness this knowledge to improve crop production, transform feed efficiency, and increase resilience to stress and disease. Emerging accounts of research on the human microbiome provide tantalizing reports of the effect of resident microbes on our body's health. In comparison, a detailed understanding of the microbiomes in agriculture—animals, plants, and soil—is markedly more rudimentary, even as their functional and critical roles have been recognized for each at a fundamental level. A better understanding of molecular-level interactions between the soil, plant, and animal microbiomes could revolutionize agriculture by improving soil structure, increasing feed efficiency and nutrient availability, and boosting resilience to stress and disease. With increasingly sophisticated tools to probe agricultural microbiomes, the next decade of research promises to bring increasing clarity to their role in agricultural productivity and resiliency.

Recommendation 5: Establish initiatives to increase the understanding of the animal, soil, and plant microbiomes and their broader applications across the food system. Transdisciplinary efforts focused on obtaining a better understanding of the various agriculturally relevant microbiomes and the complex interactions among them would create opportunities to modify and improve numerous aspects of the food and agricultural continuum. For example, understanding the microbiome in animals could help to more precisely tailor nutrient rations and increase feed efficiency. Knowing which microbes or consortia of organisms might be protective against infections could decrease disease incidence and/or severity and therefore lower losses. Research efforts are already under way to characterize the food microbiome in an effort to produce a reference database for microbes upon which rapid identification of human pathogens can be based. In plant sciences, research priorities are being established that focus on engineering various microbiomes to promote better disease control, drought resistance, and yield enhancement. Characterization of interactions between the soil and plant microbiomes is critical. The soil microbiome is responsible for cycling of carbon, nitrogen, and many other key nutrients that are required for crop productivity, and carries out several other key ecosystem functions impacted in largely unknown ways by a changing climate. Enhanced understanding of the basic microbiome components and the roles they play in nutrient cycling is likely to be critical for ensuring continuing and sustainable crop production globally.

2.4 Promising Research Directions

The committee explored important research directions to address key research challenges by disciplines or categories. Among the most promising research directions are those described in Box S-1. These are noted as important opportunities because of their potential for transforming food and agriculture and because new scientific developments make them possible in the near term. Although these research directions have focused targets, their broader objectives with respect to the research challenges are interconnected and can be pursued synergistically. The breakthrough initiatives recommended earlier would catalyze the success of the research directions in Box S-1.

BOX S-1
Recommended Research Directions by Discipline or Category

Crops

1. Continue to genetically dissect and then introduce desirable traits and remove undesirable traits from crop plants through the use of both traditional genetic approaches and targeted gene editing.
2. Enable routine genetic modification of all crop plants through the development of facile transformation and regeneration technologies.
3. Monitor plant stress and nutrients through the development of novel sensing technologies, and allow plants to better respond to environmental challenges (heat, cold, drought, flood, pests, nutrient requirements) by exploring the use of nanotechnology, synthetic biology, and the plant microbiome to develop dynamic crops that can turn certain functions on or off only when needed.

Animal Agriculture

1. Enable better disease detection and management using a data-driven approach through the development and use of sensing technologies and predictive algorithms.
2. Accelerate genetic improvement in sustainability traits (such as fertility, improved feed efficiency, welfare, and disease resistance) in livestock, poultry, and aquaculture populations through the use of big genotypic and sequence datasets linked to field phenotypes and combined with genomics, advanced reproductive technologies, and precision breeding techniques.
3. Determine objective measures of sustainability and animal welfare, how those can be incorporated into precision livestock systems, and how the social sciences can inform and translate these scientific findings to promote consumer understanding of trade-offs and enable them to make informed purchasing decisions.

continued

BOX S-1 Continued

Food Science and Technology

1. Profile and/or alter food traits for desirability (such as chemical composition, nutritional value, intentional and unintentional contamination, and quality and sensory attributes) via improvements in processing and packaging technologies, the design and functionality of sensors, and the application of "foodomic" technologies (including genomics, transcriptomics, proteomics, and metabolomics).
2. Provide enhanced product quality, nutrient retention, safety, and consumer appeal in a cost-effective and efficient manner that also reduces environmental impact and food waste by developing, optimizing, and validating advanced food processing and packaging technologies.
3. Support improved decision making to maximize food integrity, quality, safety, and traceability, as well as reduce food loss and waste by capitalizing on new data analytics, data integration, and the development of advanced decision support tools.
4. Enhance consumer understanding and acceptance of innovations in food production, processing, and safe handling of foods through expanded knowledge about consumer behavior and risk-related decisions and practices.

Soils

1. Maintain depth and quality of existing fertile soils, and restore degraded soils through adoption of best agronomic practices combined with the use of new sensing technologies, biological strategies, and integrated systems approaches.
2. Significantly increase and optimize nutrient-use efficiency (especially nitrogen) through the integration of novel sensing technologies, data analytics, precision plant breeding, and land management practices.
3. Create more productive and sustainable crop production systems by identifying and harnessing the soil microbiome's capability to produce nutrients, increase nutrient bioavailability, and improve plant resilience to environmental stress and disease.

3. FURTHER CONSIDERATIONS

The science breakthroughs alone cannot transform food and agricultural research, as there are other factors that contribute to the success of food and agricultural research. Such factors include the research infrastructure, funding, and the scientific workforce. Other considerations include the social, economic, and political outcomes of various approaches.

4. Improve the transfer of technology and practices to farmers to reduce soil loss through converging research in soil sciences, technology adoption, and community engagement.

Water-Use Efficiency and Productivity

1. Increase water-use efficiency by implementing multiple water-saving technologies across integrated systems.
2. Lower water use through applications of prescriptive analytics for water management.
3. Lower water demands by improving plant and soil properties to increase water-use efficiency.
4. Increase water productivity by use of controlled environments and alternative water sources.

Data Science

1. Accelerate innovation by building a robust digital infrastructure that houses and provides FAIR (findable, accessible, interoperable, and reuseable) and open access to agri-food datasets.
2. Develop a strategy for data science in food and agriculture research, and nurture the emerging area of agri-food informatics by adopting and influencing new developments in data science and information technology in food and agricultural research.
3. Address privacy concerns and incentivize sharing of public, private, and syndicated data across the food and agricultural enterprise by investing in anonymization, value attribution, and related technologies.

Systems Approach

1. Identify opportunities to improve the performance and adoption of integrated systems models of the food system and decision support tools.
2. Incorporate elements of systems thinking and sustainability into all aspects of the food system (from education to research to policy).

3.1 Research Infrastructure Considerations

Conclusion 1: Investments are needed for tools, equipment, facilities, and human capital to conduct cutting-edge research in food and agriculture. Addressing agriculture's most vexing problems in a coherent manner will require investments in research infrastructure that facilitate convergence of disciplines on food and agricultural research. These could include physical infrastructure for experimentation as well as cyber infrastructure that enable sharing of ideas, data, models, and knowledge. Investments in our knowledge infrastructure are needed to develop a workforce capable of working in transdisciplinary teams and in a convergent manner. Mechanisms are also needed to facilitate building private–public partnerships and engaging the public in food and agricultural research.

Conclusion 2: The Agricultural Experiment Station Network and the Cooperative Extension System deserve continued support because they are vital for basic and applied research and are needed to effectively translate research to achieve impactful results in the food and agricultural sectors. The agricultural sciences are grounded in the basic sciences but have an eye toward the applied; this has historically been facilitated by state agricultural experiment stations, as well as by extension and outreach efforts. Personnel and facilities with these functions allow scientists to translate laboratory-based findings into real-world products and processes that are most relevant, ultimately reaching key stakeholders and end users. Those stakeholders include industry, regulatory agencies, farmers and ranchers, and the general public. The recognition that scientists need to collaborate with stakeholders and translate basic research into useful and applicable results for the good of society is a fundamental value of the agricultural sciences. Recognizing and reinforcing that value through the provision of resources is essential for integrating agricultural scientific breakthroughs into the fabric of everyday life.

3.2 Funding Considerations

Conclusion 3: Current public and private funding for food and agricultural research is inadequate to address critical breakthrough areas over the next decade. There is a rapidly emerging necessity for food security and health to merit national priority and receive the funding needed to address the complex challenges in the next decade. If a robust food system is critical for securing the nation's health and well-being, then funding in both the public and private sectors ought to reflect this as a priority.

In the past century, public funding for food and agricultural research has been essential for enabling talented U.S. scientists to conduct basic scientific research and provide innovative solutions for improving food and agriculture. However, in the past decade, the United States has lost its status as the top global performer of public agricultural research and development (R&D). Unless the United States reverses this trend and invests, the United States will fall behind other countries in terms of agricultural growth. In fiscal year 2017, the National Institutes of Health (NIH) allocated $18.2 billion for competitive research grants compared to USDA, which was appropriated only $325 million for competitive research grants (less than 2 percent of the NIH's amount), a budget that was less than half of the congressionally authorized amount. The current level of federal funding for food and agricultural research has thus been inadequate. Breakthrough science needed to assist the food and agricultural enterprise to thrive in the future will require a significant investment. More will be required to sustain the level of coordination and collaboration needed to address the increasingly integrative, expansive, and visionary research required to ensure future security and competitiveness.

Although private R&D is not a substitute for public R&D funding, private foundations and industry can provide some research funding that is complementary to public funding in the U.S. agricultural innovation system. Innovative business models can be more widely employed for engaging researchers. For example, venture capital funding for start-up companies, which are well known in the tech industry, are providing record sources of investment in food and agricultural research. There are new institutions and mechanisms of financing research and of implementing innovations induced by research that offer the potential to expand funding (e.g., the Foundation for Food and Agriculture Research). However, these sources alone are insufficient to achieve the goals laid out in this report. In order for the U.S. agricultural enterprise to capitalize on the integrative, expansive, and visionary tools of research now being actively pursued by many other industries (e.g., sensing technologies, wireless communication, machine learning), a commitment to a major investment is needed now to ensure their relevant application to food and agriculture.

3.3 Education and Scientific Workforce

Conclusion 4: Efforts to renew interest in food and agriculture will need to be taken to engage non-agricultural professionals and to excite the next generation of students. Vast opportunities are available for nontraditional agricultural professionals to be involved in food and agriculture. However, there may be barriers to their involvement, such as misperceptions about the sophistication of agricultural technology and the lack of sustained funding for building transdisciplinary agricultural research teams that include non-agricultural professionals and scientists from other disciplines to work in food and agricultural sciences.

A robust workforce for food and agricultural research will require talented individuals who are proficient in the challenges facing the food system along with an understanding of the opportunities to think outside the box for innovative approaches. Recruiting talented individuals into food and agricultural research will require a demonstration and shift in perception that food and agriculture can be innovative.

3.4 Socioeconomic Contributions and Other Considerations

Conclusion 5: A better understanding of linkages between biophysical sciences and socioeconomic sciences is needed to support more effective policy design, producer adoption, and consumer acceptance of innovation in the food and agricultural sectors. The successful application of scientific innovation in the field depends on the willingness and ability of stakeholders to successfully apply and use new products and processes; it also depends on whether they view high-tech, site-specific approaches as economically or ecologically beneficial. There is a critical need to better understand the best means and methods for effective technology development and integration in production processes, with input from both the public and private sectors. Better understanding of the political economy, behavioral and choice processes related to both adoption and use of the technological innovation, and acceptance and perception of new products will be required to support the effective design of policies and application of the research innovation. For example, digital information from remote sensing devices may be inputs into a new decision support system to assist agricultural workers in making choices about field practices or animal handling. However, workers will need sufficient training and motivation to respond to expected and unexpected outcomes and uncertainty (e.g., animal response to treatment or extreme weather events). Lessons from behavioral sciences may help support behavioral change and training requirements.

The successful implementation of scientific advances also requires other important considerations to be taken into account. Policies on land or input use, environmental impact, animal welfare, and food-handling practices can have significant near- and long-term impacts on agricultural and food sustainability. Some policy or technology changes may have unintended consequences in the system and require closer examination of system interactions, including human behaviors related to adoption and use of new inputs, products, and processes. Insights from behavioral sciences can help inform the policy designs and reduce the costs of change, inform technological adoption in the field (e.g., design of conservation or tillage applications, or provision of product information to consumers), and address issues of product acceptance and consumer trust in the food system.

4. CLOSING REMARKS

At this pivotal time in history with an expanding global population requiring more from an increasingly fragile natural resource base, science breakthroughs are needed now more than ever for food and agriculture.

As the world's greatest agricultural producer, the United States bears the tremendous responsibility of implementing scientific advances to support our nation's well-being and security, and perhaps even global stability. The U.S. scientific enterprise is willing to rise to address such challenges; the tools and resources identified in this report can ensure its success.

1

Introduction

1. BACKGROUND

For nearly a century, scientific advances have fueled progress in U.S. agriculture. As one of the most productive sectors of the U.S. economy, producers have achieved dramatic increases in output with simultaneously reduced inputs (such as land, labor, and chemicals) (Wang et al., 2018). Today's farmers produce food for far more people using less land than in previous generations due to yield gains from advances in plant and animal breeding, mechanization, agricultural chemicals, and irrigation, among other improvements to agricultural production (Clancy et al., 2016). These advances have been the direct result of sustained historical investments in food and agricultural research, providing substantial social return on public investment with an estimated marginal payoff of $32.1 per dollar invested (Alston et al., 2011). Food and agricultural innovations have enabled the delivery of safe and abundant food domestically and supported a trade surplus in bulk and high-value agricultural commodities (USDA-ERS, 2018).

In the near future, the strength and responsiveness of the U.S. food and agricultural system will be tested. Recent analyses have warned that as a consequence of the growing world population, agricultural production worldwide will have collective difficulty in meeting the global demand for food and fiber (Valin et al., 2014). Achieving the higher level of productivity needed—itself a formidable task—will not be sustainable without innovative solutions to challenges posed by shortages of arable land and water, the degradation of ecosystems, and the negative impacts of climate change.

Several scientific groups have issued reports that describe these challenges in the context of the U.S. food and agricultural system and have identified opportunities to address them relative to the potential contributions of specific disciplinary or agency missions (see Box 1-1). This report builds on the opportunities identified in those reports and many others, including the White House report on *U.S. Agricultural Preparedness and the Agri-*

BOX 1-1
Highlights from Selected Recent Reports on Challenges in U.S. Food and Agriculture

USDA-ARS National Program 301 Action Plan 2018-2022 **(ARS, 2017)**

This report lays out a national program that addresses critical needs for providing crop plants with higher inherent genetic potential. The report notes the ultimate goal is in improving production efficiency, yield, sustainability, resilience, healthfulness, product quality, and value of U.S. crops. This would require continuous crop genetic improvement through more efficient and effective plant breeding. To do so includes the use of new genes and traits from the nation's gene banks, leading-edge breeding methods, data-mining, bioinformatic tools, and incisive knowledge of crop molecular and biological processes.

The Challenge of Change: Harnessing University Discovery, Engagement, and Learning to Achieve Food and Nutrition Security **(APLU, 2017)**

This report from the Association of Public & Land-Grant Universities identifies grand challenges and "pathways" in meeting food security challenges and recommends actions to meet global food needs by 2050. Themes include increasing yields while maintaining profitability and environmental sustainability; decreasing food waste; ensuring equitable food systems; and addressing the dual burdens of undernutrition and obesity. The report concludes by stating that food security should be a top priority for the nation and that transdisciplinary approaches will be needed to find solutions.

Agriculture and Applied Economics: Priorities and Solutions **(C-FARE and AAEA, 2017)**

The report identifies 10 priorities for agricultural and applied economics research and education over the next decade. Working across disciplinary boundaries, economic science related to human behavior, markets, and institution and business structures can result in new ways of using or managing plants, animals, and even the environment, transforming risks into opportunities.

BOX 1-1 Continued

Framework for a Federal Strategic Plan for Soil Science **(NSTC, 2016)**

This report from the Soil Science Interagency Working Group identifies the pressures causing soil degradation, which include population growth and movement, an increasing urban footprint, and changing demands on water and land. The report concludes that lack of a full understanding of soil ecosystem services makes it difficult to establish targets and metrics for addressing those pressures. The report calls for research to advance fundamental knowledge of soil ecosystem services and to develop ways to track soil function under changing land-use scenarios.

Phytobiomes: A Roadmap for Research and Translation **(APS, 2016)**

This report concludes that the slowing of annual yield growths for essential food crops has put the nation at a critical juncture. The report envisions that crop management would be based on systems-level knowledge of interacting components rather than management of individual components. An examination of the phytobiome would include plants, their environments, and associated organisms within the community. Examination of the phytobiome could inform plant and agroecosystem health, soil fertility, crop yields, and food quality and safety.

American Society of Animal Science (ASAS) Grand Challenges **(ASAS, 2015)**

Predicting that increases in efficiency of animal production will need to be greater during the next 40 years in order to meet the increased global demand for animal-based products, the ASAS put out five Grand Challenges in animal science. Themes include protection of human health, environmental sustainability, climate change, control of food contaminants, animal well-being, and sustainable use of water.

Unleashing a Decade of Innovation in Plant Science: A Vision for 2015-2025 **(ASPB, 2013)**

The report notes the stagnation of investment in plant-related research in the United States and the need to leverage new technologies to transform biology and accelerate the pace of discovery. The report calls for an effort to improve the ability of plant scientists to understand, predict, and alter plant behavior.

Grand Challenges for Engineering **(NAE, 2008)**

This report lays out the biggest challenges that engineers need to solve over the course of the 21st century. The list features two challenges that apply to the U.S. agriculture and food system: managing the nitrogen system and providing access to clean water. The report highlights the need for clever methods for remediating the nitrogen cycle. Agricultural irrigation consumes enormous quantities of water, often exceeding 80 percent of total water use in developing countries. This calls for improved efficiency in water use.

cultural Research Enterprise (PCAST, 2012) and the National Research Council reports *A New Biology for the 21st Century* (NRC, 2009); *Toward Sustainable Agricultural Systems in the 21st Century* (NRC, 2010); *Convergence: Facilitating Transdisciplinary Integration of Life Sciences, Physical Sciences, Engineering, and Beyond* (NRC, 2014); and *Critical Role of Animal Science Research in Food Security and Sustainability* (NRC, 2015). From the perspective of addressing the biggest problems facing the U.S. food and agricultural system, this report explores the availability of relatively new scientific tools emerging across all disciplines that could benefit the food and agricultural disciplines. This report identifies the most promising scientific breakthroughs with the potential to have the greatest impact on food and agriculture and that are possible to achieve in the next decade.

2. CHALLENGES TO THE U.S. FOOD AND AGRICULTURAL SYSTEM

2.1 Global Food System

The United States is a critical player in the global food system and is expected to be competitive in the global marketplace. A major challenge for the future is the increasing worldwide demand for food, fuel, and fiber that comes with a global population expected to reach 8.6 billion by 2030 and 9.8 billion people by 2050 (UN DESA, 2017). To meet demand in 2050, published estimates of required percentage increases in food production range from 25 to 110 percent, based partially on date of publication but also on parameters assessed, which reflect the very complex nature of the food system (Tilman et al., 2011; OECD-FAO, 2012; Ray et al., 2013; Valin et al., 2014; Hunter et al., 2017). Planning for the lower estimates could prove disastrous if the higher estimates turn out to be more accurate. Total production of meat products will need to increase worldwide by 70 percent to meet the needs of a growing global middle class with an increasing desire for animal-source foods (Robinson and Pozzi, 2011). The United States has expanded agricultural and food production and exports to help meet the demand for increased food production. Many U.S. farmers depend on export markets to expand the demand for products and support production at scales sufficient to cover costs. Similarly, the ability to export animal products, which requires freedom from reportable diseases, opens markets for U.S. production. Expanded markets for plant and animal products allow for increased scale and specialization to take advantage of new market opportunities, both of which can have a positive effect for agricultural and food producers and consumers and the overall economy. At the same time, the expanding global economy calls for increased preparation to address possible threats from abroad, including foreign animal diseases, plant pathogens, invasive pests, and trade disruptions. Increased trade and changes in the agricultural and food sector are likely to have winners and losers. It will be essential to understand the increasingly global markets, account for costs of transitions to markets and new technologies, and develop appropriate mechanisms to account for changes to support a sustainable and resilient food and agricultural system.

2.2 Plateau in Productivity

Worldwide demand for agricultural products will continue to increase, but it will become increasingly more difficult to meet those demands in the future due to an impending U.S. productivity plateau (Andersen et al., 2018). Improvements in yield potential for grain are already near their theoretical upper limit (Ort et al., 2015). Yield curves (particularly of cereal crops) are beginning to level off in some regions: approximately 30 percent of global rice, wheat, and maize production might have reached their maximum possible yields in farmers' fields (Grassini et al., 2013), and yields for rice, wheat, maize, and soybeans across 24-39 percent of the world show yields that never improve, stagnate, or collapse (Ray et al., 2012). For example, yield stagnation is occurring in the main cereal-growing areas across China, with rice yields stagnating in 53.9 percent of the counties tracked in the study, followed by 42.4 percent for maize, and 41.9 percent for wheat (Wei et al., 2015). The stagnation of productivity growth of the world's major crops (Ray et al., 2012; Grassini et al., 2013; Ort et al., 2015) serves as a warning sign that current methods for increasing crop productivity can only be exploited to a certain point, and new methods will be required to address the need for increased productivity.

2.3 Food Waste and Food Safety

On the flip side of productivity concerns is the problem of food waste. Refrigeration is considered one of the most important historical breakthroughs in agriculture because it reduces food spoilage and waste and it enhances food safety. However, further innovation to reduce and repurpose food waste is needed because the United States wastes approximately $278 billion annually, which is enough to feed nearly 260 million people (Buzby et al., 2014; Bellemare et al., 2017). Accounting for resource use and efficiencies within a systems context can help to identify opportunities for optimal waste management. Producers also need to focus on food safety, given that the Centers for Disease Control and Prevention estimates that foodborne diseases cause approximately 47.8 million illnesses, more than 125,000 hospitalizations, and about 3,000 deaths in the United States each year (Scallan et al., 2011a,b).

2.4 Resource and Environmental Constraints

Available land, water, and fertile soil are increasingly limited resources that will constrain the ability to improve productivity using today's production methods. Pushing resources past their limits risks permanent damage to the resources and to the surrounding ecosystem. The use of resource-intensive, high-input farming across the world has caused soil depletion, water scarcities, widespread deforestation, and high levels of greenhouse gas emissions (FAO, 2017). Groundwater pumping for agriculture in California has caused land sinkage, aquifer compaction, and earthquakes (Sneed and Brandt, 2015; Kraner et al., 2018). If current water use continues, the Ogallala Aquifer, which serves 30 percent of U.S. irrigation needs, will be 60 percent depleted by 2060 (Steward et al., 2013).

Although some farm practices, such as conservation tillage and no-till cropping, have been partly effective at reducing erosion (Nearing et al., 2017), soil loss continues. In 2012, an estimated 6 tons per acre of soil (roughly 1 percent of soil per acre) were lost from Iowa cropland alone, polluting rivers and streams with sediments and fertilizers. Sediment removal from waterways alone costs more than $40 billion per year (ASPB, 2013). Biotic natural resources are also at risk. The loss of native pollinators for reasons not entirely understood, together with the problems facing managed colonies of European honeybees, is estimated to cost $24 billion in lost yields annually (White House, 2014).

Global climate change adds to the challenges for the U.S. food and agricultural system. For most crops and livestock, the effects of climate change on U.S. agriculture will be increasingly negative as variable and extreme weather events, elevated temperatures, shifting rainfall patterns, prolonged dry periods, and other climate changes affect yields, with impacts varying depending on location and crop (Hatfield et al., 2014). In 2017, climate-related disasters in the United States included droughts, floods, freezes, wildfires, and hurricanes, which resulted in more than $5 billion in agricultural losses (NCEI, 2018). Climate stresses along with the recent emergence of new pests and diseases—such as citrus greening and new strains of viruses affecting swine and poultry—are imposing new demands on the nation's scientific capacities.

2.5 Changing Consumer Needs

Domestically, consumer food preferences are changing. Consumers are more acutely aware of food choices impacting their health and the environment, and large retailers are responding by distinguishing their products in the marketplace and emphasizing values such as sustainability, animal welfare, and treatment of labor in their supply chains. There is a national interest in creating market opportunities for producers to increase healthful, diverse, and affordable food choices. Consumer advocates in public health seek healthier food, citing the poor diets of Americans as one of the preventable causes of chronic disease that accounts for hundreds of billions of dollars in annual health care costs (CDC, 2017).

2.6 Declining Public Funding

The success of U.S. agriculture to date is in large part attributable to a foundation of basic and applied knowledge built in the past century by the land grant university system, the U.S. Department of Agriculture (USDA), and other federal agencies. Publicly funded research, which has produced many significant advances in agriculture, has diminished in the past 10 years due to sharp declines in the share of public investments in food and agricultural research. From 2003 to 2013, the budget for U.S. agricultural research and development (R&D) fell from $6.0 billion to $4.5 billion (Clancy et al., 2016). During the same period, private-sector funding increased by 64 percent, overtaking publicly funded research in 2010 dollars (see Figure 1-1). Both public and private funding would show even slower growth since 2009 (and even a decline for public investment) when dollars are indexed to account for the real costs of research (Heisey and Fuglie, 2018). Historically, public funding to the agricultural sector has tended to focus on advances in science and innovation, while private-sector funding has aimed at commercially useful production processes and products (Clancy et al., 2016; Heisey and Fuglie, 2018). Evidence today indicates that the two funding streams are to be viewed as complementary in the U.S. agricultural innovation system and private-sector investment does not crowd out public investment, especially in the United States (Wang et al., 2013; Fuglie and Toole, 2014; Clancy et al., 2016; Heisey and Fuglie, 2018). There is evidence, however, that public and private investments tend to go toward different sectors in research. Private research and development spending has leveraged the public research and focused on areas of commercially useful technologies that are easy to patent and protect with intellectual property protection and offer greater profit opportunities for investors. Sectors with relatively high private investment in research include food and feed manufacturing, plant systems and crop protection (especially genetically modified crops, agricultural chemicals), farm machinery and engineering, and animal health (especially veterinary pharmaceuticals) (Clancy et al., 2016).

Interest in growing consumer demand for new and diverse products, in applications of biotechnology and information sciences, and in intellectual property right protection has spurred more rapid increases in private investment (Heisey and Fuglie, 2018). However, recent mergers in some

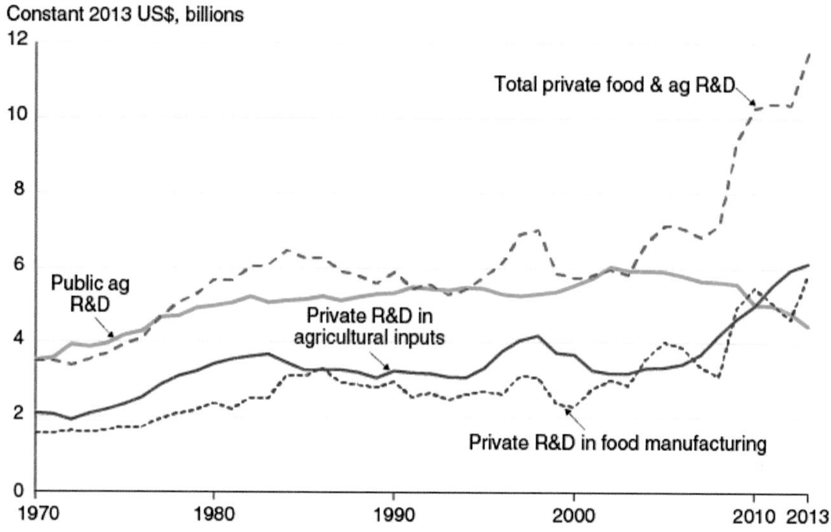

FIGURE 1-1 Annual spending on public- and private-sector research, with dollars adjusted for inflation.
NOTE: R&D = research and development.
SOURCE: Clancy et al., 2016.

agricultural industries (e.g., seed and agricultural chemicals) have led to concerns about the potential for placing farmers at a disadvantage through higher input prices, reliance on existing supply networks or technical ties to related products from the input suppliers, and for dampening incentives for the private firms to innovate (Wang et al., 2013; MacDonald, 2017). The trend of declining public research funding is concerning because it has negative implications for generating foundational research that is critical for science breakthroughs.

3. OPPORTUNITIES FOR THE FUTURE

Major scientific advances in the past decade have paved the way for new opportunities in food and agriculture. For instance, molecular biology has provided substantial advances over the past two decades that enable more precise and diverse changes in crops. These capabilities allow for the development of new food sources and traits that increase resistance to a broader array of insect pests and diseases; increase yields, nutrient-use, and water-use efficiencies; and increase the ability to withstand weather extremes. Knowledge of plants and their associated microbes at the ecosystem level (the phytobiome) also holds promise for innovation. New technologies, innovations, and insights from fields outside of mainstream agricultural disciplines are also empowering producers. In 2016, farmers accounted for approximately 19 percent of commercially used unmanned aerial vehicles (also known as drones), because they are seeking cost-effective ways to identify disease and weeds and determine field conditions (Hogan et al., 2017; Hunt and Daughtry, 2017). Nanotechnology offers improved capabilities to (1) sense and monitor physical, chemical, or biological properties and processes (to ultimately improve the sustainability of food production); (2) control microbes (to improve food safety and minimize food waste); and (3) create new materials (to monitor and improve animal health) (Rodrigues et al., 2017). Poultry producers are implementing computerized approaches such as artificial intelligence to fine-tune their strategies for feeding chickens and monitoring their health while reducing labor and feeding costs (Ahmed, 2011; Hepworth et al., 2012). Certified-organic lettuce is being grown in soil-free systems in urban and peri-urban environments (Dewey, 2017), and the number of local farmers' markets across the nation continues to grow (Low et al., 2015). Food and agricultural research advances will need to integrate innovations to simultaneously address water scarcity, soil health, food waste, pests and diseases, climate variability, and overall system sustainability.

The research directions described in this report depend on assimilating cutting-edge developments from allied fields—such as computing, information science, machine learning, materials science and electronics, genomics and gene editing, and behavioral and cognitive science—to achieve solutions to overarching and complex problems faced by agriculture. Leveraging the advances from other disciplines implies a need for the food and agricultural sciences to attract and train research talent in those areas. Some of the research approaches identified focus on systems-level discovery that will require regional or national cooperation and planning, in addition to a multidisciplinary effort. If successful, these systems approaches will produce essential information that will be the basis for new knowledge to inform decision making at different scales and tools to implement those decisions.

4. PURPOSE OF THIS STUDY

The trends described in the foregoing section set the stage for the interest of the Supporters of Agricultural Research (SoAR) Foundation, in partnership with the Foundation for Food and Agriculture Research (FFAR), along with professional societies, commodity groups, and farmer organizations to propose the need for a strategic vision from the agricultural science community that articulates the greatest opportunities within the field, and the potential pathways that will lead to a new generation of scientific

BOX 1-2
Statement of Task

An executive committee, assisted by science panels, will be appointed to lead the development of an innovative strategy for the future of food and agricultural research, answering the following questions:

1. What are the greatest challenges that food and agriculture are likely to face in the coming decades?
2. What are the greatest foreseeable opportunities for advances in food and agricultural science?
3. What fundamental knowledge gaps exist that limit the ability of scientists to respond to these challenges as well as take advantage of the opportunities?
4. What general areas of research should be advanced and supported to fill these knowledge gaps?

In the process of addressing these questions, the committee will gather insights from scientists and engineers in the traditional fields of science in food and agriculture, seek ideas from scientists in other disciplines whose knowledge, tools, and techniques might be applied to food and agricultural challenges, and organize interdisciplinary dialogues to uncover novel, potentially transformational, approaches to advancing food and agricultural science.

At the end of its exploration, the executive committee will produce a consensus report recommending future research directions in food and agriculture. The committee will frame its recommendations in the context of the importance and relevance of the science to the public's interest in the benefits of catalyzing knowledge creation—a sustainable food and fiber supply, better public health, a strengthened natural resource base, and the creation of new economic opportunities and jobs. Prior to its finalization, the report will be anonymously peer reviewed, and revised by the executive committee before release to the sponsors and the public.

advancement. The USDA's National Institute of Food and Agriculture, the National Science Foundation, and the U.S. Department of Energy all agreed to support an endeavor of this nature in conjunction with the SoAR Foundation, FFAR, and their other partners. There has been a collective interest in an exercise that might begin a tradition of a "decadal survey" for food and agriculture in laying out research priorities, a long-standing practice used in some fields of science to guide the programmatic focus of federal research agencies in 10-year increments. With decadal surveys being instrumental to fields such as space studies in prioritizing research needs for the next 10 years, the intent is that a study such as this would be useful in informing strategic planning and discussions on food and agricultural research.

Responding to the call for such a study, the National Academies of Sciences, Engineering, and Medicine (the National Academies) convened an ad hoc study committee to provide a broad new vision for food and agricultural research by outlining the most promising science breakthroughs over the next decade (see the Statement of Task in Box 1-2).

5. APPROACH TO THE TASK

5.1 Committee and Information-Gathering Meetings

The National Academies convened a committee of 13 experts with collective expertise and experience in various disciplines. The collective expertise of the committee allowed it to address plants, animals, microbes, food science, food safety, human nutrition, soil, water, climate, ecology, pests, and pathogens, as well as landscape and/or watershed systems, agricultural economics, transdisciplinary fields (sustainability, biodiversity), and emerging technological applications at the frontiers of agriculture (nanotechnology, biotechnology, remote sensing, data mining, machine learning, modeling, robotics) (see Appendix A for the biographical sketches of the committee members).

The committee held several meetings and webinars as part of the information-gathering process (see Appendix B for the open session meeting agendas). A broad solicitation was initially sent to scientific societies and stakeholder groups asking individuals to identify research that would advance science and promote solutions and opportunities in food and agriculture, with the responses viewable on IdeaBuzz, an interactive online discussion platform (see Appendix C for the IdeaBuzz submission contributors and a summary of the ideas submitted). Based on some of those responses, the committee identified focal areas of research and invited a large panel of experts to assist the committee at a 3-day public meeting in describing major opportunities to advance science, identify knowledge gaps, prioritize research barriers to overcome, and articulate a strategy for moving forward (see Appendix B for the "Jamboree" agenda and meeting participants). It was recognized early on that a transdisciplinary approach was needed to address the complexities and interdependencies of the food and agricultural system; thus, a diverse community of scientists was invited from both the traditional agricultural disciplines and allied fields.

As part of its charge from the sponsors, the following criteria were provided to assist the committee in identifying the most promising scientific breakthroughs in food and agriculture:

1. An emphasis on identifying transformational research opportunities to address key challenges in food and agriculture;
2. Recognition that the complexity of most food and agriculture challenges requires transdisciplinary and integrated approaches to the development of lasting solutions;
3. The value of harnessing insights from the frontiers of scientific disciplines and communities not traditionally associated with food and agriculture;
4. The importance of involving stakeholders of all kinds in the process and of raising public awareness of the meaning and significance of this scientific discussion; and
5. A compressed time frame (relative to typical decadal surveys) to draw together diverse communities, explore ideas, analyze proposed goals, and produce consensus recommendations for the strategy.

5.2 Scope of the Charge

The committee recognized early on that it had to place limitations on the scope of its study. The food and agricultural system encompasses everything from production through processing and distribution, and it would have been impossible to examine and address all aspects of the system in this report. Instead, the committee focused its efforts on parts of the system that require attention and are challenging yet hold the most promising opportunities for scientific breakthroughs in the near future. This report is not intended to provide a roadmap to span the entire spectrum of food and agricultural research, but rather to suggest a strategy to capitalize on several key potential science breakthroughs for transformational change.

By identifying the most challenging issues in food and agriculture, the committee delineated boundaries for the study and determined that certain areas were outside the scope of consideration for science breakthroughs. These include topics such as biodiversity, biofuels, food distribution and equity, food access and insecurity, and fundamental human nutrition science. For areas such as human nutrition and obesity, the committee did not attempt to address those topics because other groups have already described strategies and roadmaps for nutrition research (ICHNR, 2016); however, the committee acknowledges that a diverse, nutritious food supply is an integral goal of the food and agricultural system. Also, the committee recognizes that the U.S. food system exists in a global context, but limited its focus to U.S. issues for determining priority areas and targeting its recommendations to U.S. researchers, policy makers, and stakeholders.

6. GOALS FOR 2030

In looking toward the next decade, the committee identified major goals for food and agricultural research to include (1) improving the efficiency of food and agricultural systems, (2) increasing the resiliency of agricultural systems to adapt to rapid changes and extreme conditions, and (3) increasing the sustainability of agriculture. Efficiency (specifically, technical and allocative efficiency) refers to the ability to obtain a maximum level of output from available inputs (resources such as energy, water, labor, and capital) and at lowest possible cost (Wang et al., 2015; Shumway et al., 2016). Resilience is defined as "the ability to prepare and plan for, absorb, recover from, and successfully adapt to adverse events" (NRC, 2012; p. 1). It may refer to the viability and adaptability of individual plant and animal species that are cultivated and consumed by human populations, and it may also describe certain properties of the food system as a whole. Sustainability refers to the ability to meet society's need for food without compromising the ability of future generations to meet their needs (UC Davis, 2018). More specifically, USDA defines sustainable agriculture as

> an integrated system of plant and animal production practices having a site-specific application that will over the long-term: (1) satisfy human food and fiber needs; (2) enhance environmental quality and the natural resource base upon which the agriculture economy depends; (3) make the most efficient use of nonrenewable resources and on-farm resources and integrate, where appropriate, natural biological cycles and controls; (4) sustain the economic viability of farm operations; and (5) enhance the quality of life for farmers and society as a whole. (U.S. Code Title 7, Section 3103)

Integrating approaches and technologies across various disciplines is essential, with the ultimate goal of improving the quality and increasing the quantity of food to sustainably meet our needs.

To achieve the major goals of efficiency, resiliency, and sustainability, improvements are needed to address the most challenging issues across the food system. The most challenging issues were derived from the common nature of important research challenges identified by food and agricultural scientists and reiterated by the committee, and include the following:

- increasing nutrient use efficiency in crop production systems;
- reducing soil loss and degradation;
- mobilizing genetic diversity for crop improvement;
- optimizing water use in agriculture;
- improving food animal genetics;
- developing precision livestock production systems;

- early and rapid detection and prevention of plant and animal diseases;
- early and rapid detection of foodborne pathogens; and
- reducing food loss and waste throughout the supply chain.

7. ORGANIZATION OF THE REPORT

This report identifies opportunities to boost the performance of the U.S. food and agricultural system, reduce its impact on the environment, help it expand in new directions, and increase its resilience in the face of environmental uncertainty. Based on the previously noted challenge areas, the committee explores seven specific areas in which such advances could be made:

1. **Crops** (Chapter 2)—The advent of more precise gene-editing technologies opens new avenues for achieving the goals of increasing crop productivity while decreasing inputs and improving resilience. The chapter discusses the need for new traits, facile transformation technology, and dynamic crops where responses to environmental challenges can be turned on or off.
2. **Animal Agriculture** (Chapter 3)—Decades of R&D have dramatically improved the efficiency of animal production over the past century, but additional investment is critical to sustainably address the expected twofold increase in animal products. The chapter also examines issues related to animal health well-being and animal-source food alternatives.
3. **Food Science and Technology** (Chapter 4)—The postharvest food sector ensures that raw agricultural products are converted to a safe, nutritious, sustainable, and affordable food supply that is readily available to all. This chapter examines issues related to protecting and enhancing food quality, safety, and appeal while simultaneously reducing food loss and waste.
4. **Soils** (Chapter 5)—Maintaining and properly managing fertile soils is a critical need to ensure agricultural productivity. This chapter discusses soil sustainability, soil quality, nutrient availability, and the soil microbiome.
5. **Water-Use Efficiency and Productivity** (Chapter 6)—Fresh water is a finite resource. Meeting increasing demands for food, fuel, and fiber can only be accomplished with increased water efficiency. This chapter examines opportunities to improve the use of data analytics, improve plant and soil properties, and capitalize on the use of controlled environments for agriculture.

6. **Data Science** (Chapter 7)—Data science can be better harnessed to improve aspects of the food system. This chapter examines the opportunities and advancements in data science and information technologies on the horizon for the food and agricultural sectors.
7. **A Systems Approach** (Chapter 8)—The food system operates in the context of a complex system with many actors and components. This chapter examines the need for better understanding of the various systems and components as they relate to a functional food and agricultural enterprise.

The final chapter (Chapter 9) presents the strategy for 2030 along with the five breakthrough areas and the overarching study recommendations. Chapter 9 also discusses crosscutting issues for future consideration, including research infrastructure, societal dynamics, and education and workforce needs.

REFERENCES

Ahmed, H. 2011. Egg production forecasting: Determining efficient modeling approaches. *Journal of Applied Poultry Research* 20(4):463-473.

Alston, J. M., M. A. Andersen, J. S. James, and P. G. Pardey. 2011. The economic returns to U.S. public agricultural research. *American Journal of Agricultural Economics* 93(5):1257-1277.

Andersen, M. A., J. M. Alston, P. G. Pardey, and A. Smith. 2018. A century of U.S. farm productivity growth: A surge then a slowdown. *American Journal of Agricultural Economics* 1-19. doi: 10.1093/ajae/aay023.

APLU (Association of Public & Land-Grant Universities). 2017. *The Challenge of Change: Harnessing University Discovery, Engagement, and Learning to Achieve Food and Nutrition Security.* Washington, DC: APLU. Available at http://www.aplu.org/library/the-challenge-of-change/file (accessed May 8, 2018).

APS (American Phytopathological Society). 2016. *Phytobiomes: A Roadmap for Research and Translation.* St. Paul, MN: American Phytopathological Society. Available at https://www.apsnet.org/members/outreach/ppb/Documents/PhytobiomesRoadmap.pdf (accessed May 8, 2018).

ARS (Agricultural Research Service). 2017. *National Program 301: Plant Genetic Resources, Genomics and Genetic Improvement: Action Plan 2018-2022.* Available at https://www.ars.usda.gov/ARSUserFiles/np301/NP%20301%20Action%20Plan%202018-2022%20FINAL.pdf (accessed May 8, 2018).

ASAS (American Society of Animal Science). 2015. *ASAS Grand Challenges.* Available at https://www.asas.org/about/public-policy/asas-grand-challenges (accessed May 8, 2018).

ASPB (American Society of Plant Biologists). 2013. *Unleashing a Decade of Innovation in Plant Science: A Vision for 2015-2025.* Available at https://plantsummit.files.wordpress.com/2013/07/plantsciencedecadalvision10-18-13.pdf (accessed May 8, 2018).

Bellemare, M. F., M. Cakir, H. H. Peterson, L. Novak, and J. Rudi. 2017. On the measurement of food waste. *American Journal of Agricultural Economics* 99(5):1148-1158.

Buzby, J. C., F. W. Hodan, and J. Hyman. 2014. *The Estimated Amount, Value, and Calories of Postharvest Food Losses at the Retail and Consumer Levels in the United States*. Economic Information Bulletin No. EIB-121. Washington, DC: USDA Economic Research Service. Available at https://www.ers.usda.gov/publications/pub-details/?pubid=43836 (accessed May 10, 2018).

CDC (Centers for Disease Control and Prevention). 2017. *Chronic Disease Overview*. Available at https://www.cdc.gov/chronicdisease/overview/index.htm (accessed July 3, 2018).

C-FARE and AAEA (Council on Food, Agricultural and Resource Economics and Agricultural and Applied Economics Association). 2017. *Agricultural and Applied Economics Priorities and Solutions Report*. Available at https://static1.squarespace.com/static/598b4450e58c624720903ae6/t/59a76b8012abd9e692d6d623/1504144269634/PrioritiesandSolutionsReport04-06-2017-LOW_v2%282%29.pdf (accessed July 3, 2018).

Clancy, M., K. Fuglie, and P. Heisey. 2016. *U.S. Agricultural R&D in an Era of Falling Public Funding*. U.S. Department of Agriculture, Economic Research Service. Available at https://www.ers.usda.gov/amber-waves/2016/november/us-agricultural-rd-in-an-era-of-falling-public-funding (accessed May 4, 2018).

Dewey, C. 2017. Pioneers of organic farming are threatening to leave the program they helped create. *The Washington Post*. November 2. Available at https://www.washingtonpost.com/news/wonk/wp/2017/11/02/pioneers-of-organic-farming-are-threatening-to-leave-the-program-they-helped-create/?noredirect=on&utm_term=.ddd4e948a7e3 (accessed May 11, 2018).

FAO (Food and Agriculture Organization of the United Nations). 2017. *The Future of Food and Agriculture: Trends and Challenges*. Available at http://www.fao.org/3/a-i6583e.pdf (accessed June 13, 2018).

Fuglie, K. O., and A. A. Toole. 2014. The evolving institutional structure of public and private agricultural research. *American Journal of Agricultural Economics* 96(3):862-883.

Grassini, P., K. M. Eskridge, and K. G. Cassman. 2013. Distinguishing between yield advances and yield plateaus in historical crop production trends. *Nature Communications* 4:2918.

Hatfield, J., G. Takle, R. Grotjahn, P. Holden, R. C. Izaurralde, T. Mader, E. Marshall, and D. Liverman. 2014. Ch. 6: Agriculture. *Climate Change Impacts in the United States: The Third National Climate Assessment,* J. M. Melillo, Terese (T.C.) Richmond, and G. W. Yohe, Eds., U.S. Global Change Research Program, 150-174. doi: 10.7930/J02Z13FR.

Heisey, P. W., and K. O. Fuglie. 2018. *Agricultural Research Investment and Policy Reform in High-Income Countries*. Available at https://www.ers.usda.gov/webdocs/publications/89114/err-249.pdf?v=43244 (accessed June 11, 2018).

Hepworth, P. J., A. V. Nefedov, I. B. Muchnik, and K. L. Morgan. 2012. Broiler chickens can benefit from machine learning: Support vector machine analysis of observational data. *Journal of the Royal Society, Interface* 9(73):1934-1942.

Hogan, S., M. Kelly, B. Stark, and Y. Chen. 2017. Unmanned aerial systems for agriculture and natural resources. *California Agriculture* 71(1):5-14.

Hunt, E. R., Jr., and C. S. T. Daughtry. 2017. What good are unmanned aircraft systems for agricultural remote sensing and precision agriculture? *International Journal of Remote Sensing*. Available at https://doi.org/10.1080/01431161.2017.1410300 (accessed July 3, 2018).

Hunter, M. C., R. G. Smith, M. E. Schipanski, L. W. Atwood, and D. A. Mortenson. 2017. Agriculture in 2050: Recalibrating targets for sustainable intensification. *BioScience* 67:386-391.

ICHNR (Interagency Committee on Human Nutrition Research). 2016. *National Nutrition Research Roadmap 2016-2021: Advancing Nutrition Research to Improve and Sustain Health*. Available at https://www.nal.usda.gov/sites/default/files/fnic_uploads/2016-03-30-%20ICHNR%20NNRR%20%282%29.pdf (accessed June 26, 2018).

Kraner, M. L., W. E. Holt, and A. A. Borsa. 2018. Seasonal non-tectonic loading inferred from cGPS as a potential trigger for the M6.0 South Napa earthquake. *Journal of Geophysical Research*. Available at https://doi.org/10.1029/2017JB015420 (accessed June 11, 2018).

Low, S. A., A. Adalja, E. Beaulieu, N. Key, S. Martinez, A. Melton, A. Perez, K. Ralston, H. Stewart, S. Suttles, S. Vogel, and B. B. R. Jablonski. 2015. *Trends in U.S. Local and Regional Food Systems: A Report to Congress*. Administrative Publication No. AP-068 U.S. Department of Agriculture, Economic Research Service, January. Available at https://www.ers.usda.gov/webdocs/publications/42805/51173_ap068.pdf?v=42083 (accessed June 11, 2018).

MacDonald, J. 2017. Mergers and Competition in Seed and Agricultural Chemical Markets. *Amber Waves*, US Department of Agriculture, Economic Research Service. April 3. Available at https://www.ers.usda.gov/amber-waves/2017/april/mergers-and-competition-in-seed-and-agricultural-chemical-markets (accessed June 11, 2018).

NAE (National Academy of Engineering). 2008. *NAE Grand Challenges for Engineering. Updated 2017*. Available at https://www.engineeringchallenges.org (accessed June 11, 2018).

NCEI (National Centers for Environmental Information). 2018. *U.S. Billion-Dollar Weather and Climate Disasters: Table of Events*. Available at https://www.ncdc.noaa.gov/billions/events/US/2017-2018 (accessed May 9, 2018).

Nearing, M. A., Y. Xie, B. Liu, and Y. Ye. 2017. Natural and anthropogenic rates of soil erosion. *International Soil and Water Conservation Research* 5(2):77-84.

NRC (National Research Council). 2009. *A New Biology for the 21st Century*. Washington, DC: The National Academies Press.

NRC. 2010. *Toward Sustainable Agricultural Systems in the 21st Century*. Washington, DC: The National Academies Press.

NRC. 2012. *Disaster Resilience: A National Imperative*. Washington, DC: The National Academies Press.

NRC. 2014. *Convergence: Facilitating Transdisciplinary Integration of Life Sciences, Physical Sciences, Engineering, and Beyond*. Washington, DC: The National Academies Press.

NRC. 2015. *Critical Role of Animal Science Research in Food Security and Sustainability*. Washington, DC: The National Academies Press.

NSTC (National Science and Technology Council). 2016. *The State and Future of U.S. Soils: Framework for a Federal Strategic Plan for Soil Science*. Available at https://obamawhitehouse.archives.gov/sites/default/files/microsites/ostp/ssiwg_framework_december_2016.pdf (accessed June 11, 2018).

OECD-FAO (Organisation for Economic Co-operation and Development and the Food and Agricultural Organization of the United Nations). 2012. *OECD-FAO Agricultural Outlook 2012*, OECD Publishing. Available at https://doi.org/10.1787/19991142 (accessed July 19, 2018).

Ort, D. R., S. S. Merchant, J. Alric, A. Barkan, R. E. Blankenship, R. Bock, R. Croce, M. R. Hanson, J. M. Hibberd, S. P. Long, T. A. Moore, J. Moroney, K. K. Niyogi, M. A. J. Parry, P. P. Peralta-Yahya, R. C. Prince, K. E. Redding, M. H. Spalding, K. J. van Wijk, W. F. J. Vermaas, S. von Caemmerer, A. P. M. Weber, T. O. Yeates, J. S. Yuan, and X. G. Zhu. 2015. Redesigning photosynthesis to sustainably meet global food and bioenergy demand. *Proceedings of the National Academy of Sciences of the United States of America* 112(28):8529-8536.

PCAST (President's Council of Advisors on Science and Technology). 2012. *U.S. Agricultural Preparedness and the Agricultural Research Enterprise*.

Ray, D. K., N. Ramankutty, N. D. Mueller, P. C. West, and J. A. Foley. 2012. Recent patterns of crop yield growth and stagnation. *Nature Communications* 3:1293.

Ray, D. K., N. D. Mueller, P. C. West, and J. A. Foley. 2013. Yield trends are insufficient to double global crop production by 2050. *PLoS ONE* 8(6):e66428. Available at https://doi.org/10.1371/journal.pone.0066428 (accessed June 11, 2018).

Robinson, T. P., and F. Pozzi. 2011. *Mapping Supply and Demand for Animal-Source Foods to 2030*. Animal Production and Health Working Paper No. 2. Rome: Food and Agriculture Organization of the United Nations. Available at http://www.fao.org/docrep/014/i2425e/i2425e00.pdf (accessed June 11, 2018).

Rodrigues, S. M., P. Demokritou, N. Dokoozlian, C. Ogilvie Hendren, B. Karn, M. S. Mauter, O. A. Sadik, M. Safarpour, J. M. Unrine, J. Viers, P. Welle, J. C. White, M. R. Wiesnerde, and G. V. Lowry. 2017. Nanotechnology for sustainable food production: Promising opportunities and scientific challenges. *Environmental Science: Nano* 4(4):767-781.

Scallan, E., R. M. Hoekstra, F. J. Angulo, R. V. Tauxe, M.-A. Widdowson, S. L. Roy, J. L. Jones, and P. M. Griffin. 2011a. Foodborne illness acquired in the United States—major pathogens. *Emerging Infectious Diseases* 17(1):7-15.

Scallan, E., P. M. Griffin, F. J. Angulo, R. V. Tauxe, and R. M. Hoekstra. 2011b. Foodborne illness acquired in the United States—unspecified agents. *Emerging Infectious Diseases* 17(1):16-22.

Shumway, C. R., B. M. Fraumeni, L. E. Fulginiti, J. D. Samuels, and S. E. Stefanou. 2016. U.S. Agricultural Productivity: A Review of USDA Economic Research Service Methods. *Applied Economic Perspectives and Policy* 38(1):1-29.

Sneed, M., and J. T. Brandt. 2015. Land subsidence in the San Joaquin Valley, California, USA, 2007–2014. *Proceedings of the International Association of Hydrological Sciences* 372:23-27.

Steward, D. R., P. J. Bruss, X. Yang, S. A. Staggenborg, S. M. Welch, and M. D. Apley. 2013. Tapping unsustainable groundwater stores for agricultural production in the High Plains Aquifer of Kansas, projections to 2110. *Proceedings of the National Academy of Sciences of the United States of America* 110(37):E3477-E3486.

Tilman, D., C. Balzer, J. Hill, and B. L. Befort. 2011. Global food demand and the sustainable intensification of agriculture. *Proceedings of the National Academy of Sciences of the United States of America* 108(50):20260-20264.

UC Davis (University of California, Davis). 2018. *What Is Sustainable Agriculture?* Available at http://asi.ucdavis.edu/programs/sarep/about/what-is-sustainable-agriculture (accessed June 20, 2018).

UN DESA (United Nations, Department of Economic and Social Affairs, Population Division). 2017. *World Population Prospects: The 2017 Revision, Key Findings and Advance Tables*. ESA/P/WP/248. Available at https://esa.un.org/unpd/wpp/Publications/Files/WPP2017_KeyFindings.pdf (accessed June 20, 2018).

USDA-ERS (U.S. Department of Agriculture's Economic Research Service). 2018. U.S. Agricultural Trade at a Glance. Available at https://www.ers.usda.gov/topics/international-markets-us-trade/us-agricultural-trade/us-agricultural-trade-at-a-glance (accessed June 20, 2018).

Valin, H., R. D. Sands, D. van der Mensbrugghe, G. C. Nelson, H. Ahammad, E. Blanc, B. Bodirsky, S. Fujimori, T. Hasegawa, P. Havlik, E. Heyhoe, P. Kyle, D. Mason-D'Croz, S. Paltsev, S. Rolinski, A. Tabeau, H. van Meijl, M. von Lampe, and D. Willenbockel. 2014. The future of food demand: Understanding differences in global economic models. *Agricultural Economics* 45:51-67.

Wang, S. L., P. W. Heisey, W. E. Huffman, and K. O. Fuglie. 2013. Public R&D, private R&D, and U.S. agricultural productivity growth: Dynamic and long-run relationships. *American Journal of Agricultural Economics* 95(5):1287-1293.

Wang, S. L., P. Heisey, D. Schimmelpfenning, and E. Ball. 2015. *Agricultural Productivity Growth in the United States: Measurements, Trends, and Drivers*. Economic Research Report No. (ERR0189). 78 pp. Available at https://www.ers.usda.gov/publications/pub-details/?pubid=45390 (accessed June 22, 2018).

Wang, S. L., R. Nehring, and R. Mosheim. 2018. *Agricultural Productivity Growth in the United States: 1948-2015*, ERS. Available at https://www.ers.usda.gov/amber-waves/2018/march/agricultural-productivity-growth-in-the-united-states-1948-2015/ (accessed June 20, 2018).

Wei, X., Z. Zhang, P. Shi, P. Wang, Y. Chen, X. Song, and F. Tao. 2015. Is yield increase sufficient to achieve food security in China? *PLoS ONE* 10(2):e0116430. Available at http://doi.org/10.1371/journal.pone.0116430 (accessed June 20, 2018).

White House, Office of the Press Secretary. 2014. Fact Sheet: The Economic Challenge Posed by Declining Pollinator Populations. June 20, 2014.

2

Crops

1. INTRODUCTION

Crop agriculture is one of America's success stories. U.S. farmers produce 34.1 percent of the world's soybeans, 35.5 percent of the world's corn, 13.4 percent of the world's cotton, and 7.6 percent of the world's wheat (USDA-FAS, 2018b). Crop production accounts for approximately $194 billion per year in agricultural cash receipts (USDA-ERS, 2018). In 2016, agricultural domestic exports of crops reached $108 billion ($21 billion in soybeans alone) and created an estimated $171.3 billion in additional economic activity (USTR, 2017; USDA-FAS, 2018a).

Yields of major staple crops grown in the United States are the highest or near to the highest in the world. Since the 1930s, yields of corn have increased more than eightfold, while soybean and cotton yields have increased more than fourfold, due primarily to advances in plant breeding, as well as fertilizer use and equipment efficiency, among other innovations (Fernandez-Cornejo, 2004; Nielsen, 2017). Despite successive years of weather disasters, the yield performance of America's premier crops is a remarkable testament to the resilience and success of the varieties commercially available today (see Figure 2-1).

With increased support for public-sector agricultural research and public–private partnerships, it will be possible to bring the breeding success in corn to many other crops (such as cover crops, fruits and vegetables, and bioenergy crops) at a much faster pace than is currently possible. Moreover, these efforts can be used to "harden" crops against the effects of extreme weather and increased pest and disease pressure while maximizing yield and

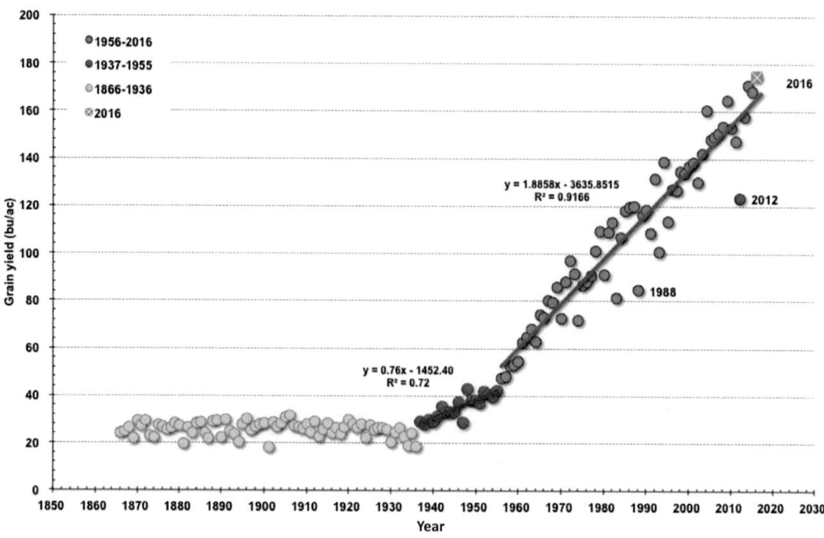

FIGURE 2-1 Yield trends for corn in the United States from 1866 to 2015.
SOURCE: Nielsen, 2017.

increasing nutritional content and flavor. These efforts can also be used to enhance other crop characteristics, such as those that will reduce the need for costly inputs. As stationary organisms, plants exhibit enormous plasticity in the face of adverse and extreme environments. Through the process of natural selection, plant communities adapt to local conditions and provide for their sustenance by interacting with their environments both below and above ground. Plants have evolved to grow and thrive in almost all environments on this planet—from lands with virtually no water to lands that flood routinely, and in all temperature extremes. This evolutionary pressure has given rise to the incredible genetic diversity of plant life that, when coupled with new gene-editing technologies, offers exciting new avenues for solving some of the big challenges facing crop production today.

Other chapters in this report will address soil, fertilizer use (see Chapter 5), and water (see Chapter 6), proper management of which is critical for sustainable, environmentally responsible crop production. There is a pressing need for changes in agronomic practices that improve the stewardship of natural resources and decrease the environmental impacts of crop production. The focus of this chapter is on the crop toolbox and how crop germplasm can be developed over the next two decades—efforts that can be pursued in conjunction with agronomy and crop science to facilitate increased diversity and greater flexibility in cropping systems. The domestication of wild plant species suited to different environments has commercial value in crop rotations and these species can serve as habitats for beneficial insects, with the potential to expand options for crop management. Similarly, the genetic tools are rapidly becoming available to endow current cultivars with traits that minimize the need for inputs (e.g., water, nitrogen, phosphate, pesticides, and fungicides), maximize yield in changing environments/climatic extremes, resist diseases and pests, and provide improved nutrition, among other characteristics.

2. CHALLENGES

It is no longer safe to assume that U.S. crop production is in a self-sustaining, steady state. First and foremost, the natural resource base for U.S. agriculture is recognized as increasingly fragile. Groundwater and fertile soil are finite resources, and their use and misuse define the boundaries of sustainable production in the long term (Rockström et al., 2017). Aquifers supporting the majority of U.S. production are being drained (Konikow, 2013; Steward et al., 2013), and soil quality in some parts of the country is degrading (Baumhardt et al., 2015). These developments set the stage for lower productivity and the need for crops that can perform well in less than optimal environments.

Second, crop systems are also stressed by changing weather patterns and extreme weather events (Walthall et al., 2012). Prolonged drought over the past decade (Diffenbaugh et al., 2015; Howitt et al., 2015) and flooding more recently (Mallakpour and Villarini, 2015) have been responsible for the largest proportion of U.S. crop disaster payments from abiotic and biotic stresses (see Figure 2-2). The need for crops that are resilient to multiple abiotic stresses, such as drought and flooding, will be a challenging problem for breeders over the next decade. There are success stories in addressing some of the challenges, such as the development of flood-tolerant varieties of rice via introduction of the *SUB1A* gene into modern cultivars (Bailey-Serres et al., 2012; see Box 2-1).

Third, biotic challenges to crops including pests and diseases are an increasing threat due to agricultural intensification, expanding global trade, and extreme weather. A recent example is Huanglongbing (HLB), also known as citrus greening disease. HLB was detected in Asia more than 100 years ago and was first seen in the United States in Florida orchards in 2005. Within 3 years the vector-transmitted bacterial pathogen spread to the majority of citrus orchards in Florida, leading to losses of more than $1 billion annually (Court et al., 2017). There is no known cure for HLB, which kills trees in 3-5 years and has now spread to California, Georgia, Hawaii, Louisiana, South Carolina, and Texas (NASEM, 2018).

In another example, rising temperatures are leading to migrations of

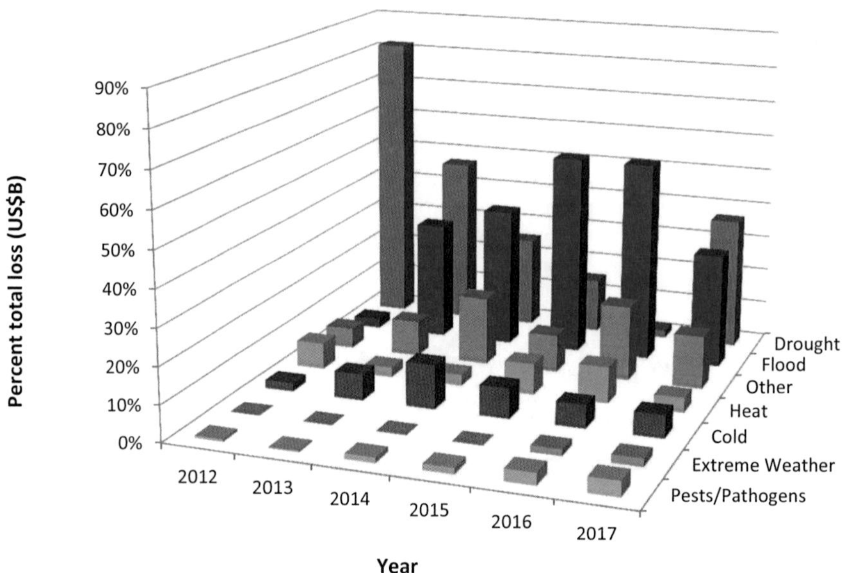

FIGURE 2-2 Percentage of total annual crop losses due to different environmental stresses.
SOURCE: USDA-RMA, 2018.

BOX 2-1
All Major Crops Except Rice Are Flooding Intolerant

The introduction of a gene called *SUB1A* from an Indian farmer's rice variety into modern rice cultivars proved sufficient to increase tolerance of full submergence to 2 weeks or more. *SUB1A* rice varieties bred using selection with molecular markers are now grown by millions of farmers. Rice survives root waterlogging because of traits that enable gases to be exchanged between the root and shoot. In the United States, extreme rains cause prolonged waterlogging of the root systems of soybean, maize, wheat, and other crops, resulting in more than $1 billion in crop loss per year. The improvement of crop flooding resilience might be accomplished through breeding, genetic engineering, or other strategies. The genetic loci of crops or their relatives that enhance waterlogging tolerance can be recognized through advanced phenotyping and mapping of diverse germplasm. But because a plant may experience both flooding and drought during its life cycle, it is critical to combine resilience to both extremes of water availability in a single seed (Bailey-Serres et al., 2012).

insect pests to higher elevations, while greater intensification is leading to more rapid evolution and spread of diseases (Katsaruware-Chapoto et al., 2017). Direct yield losses caused by pathogens and pests are predicted to lower global agricultural productivity in the future by 20 to 40 percent (Savary et al., 2012; Baltes et al., 2017). Some natural threats to crop production have been made worse by management practices, such as the emergence of herbicide-resistant weeds following overuse of a single herbicide (e.g., glyphosate) to control weeds. Improved resistance management practices can reduce input costs, improve yields, and increase returns (Livingston et al., 2015).

There is a deep sense of urgency and opportunity to speed up and expand the power of crop improvement to address these challenges. As previously mentioned, global food production will need to increase by at least 50 percent by 2030 to feed a growing global population (UN DESA, 2017). However, the current pace of yield growth worldwide may not be sufficient to meet the predicted need for 2030 and beyond (Ray et al., 2012), and there are even signs that the rate of yield increase in many crops is beginning to plateau (Wei et al., 2015; Andersen et al., 2018). The question of why yield growth is slowing has yet to be answered, with speculation ranging from the decline in agricultural research for basic crop improvement (Pardey and Beddow, 2013; Andersen, 2015), temperature and other effects of climate change (Lobell and Field, 2007), or that some crops are nearing their biological yield potential (Evans and Fischer, 1999; Alston et al., 2009; Ort et al., 2015).

3. OPPORTUNITIES

3.1 Gene-Editing Systems

The discovery of gene-editing systems (such as CRISPR-Cas9) has revolutionized our ability to both understand and genetically modify both plants and animals (Yin et al., 2017). For crops, new alleles can be generated and introduced directly into a cultivar of choice, leaping over the time-consuming process of making multiple crosses to combine desirable traits in the progeny. Gene editing creates the potential to identify and implement new traits in the field on a much faster timescale. Traditional plant breeding is slow and tedious as it can only exploit the limited quantitative trait alleles found in wild relatives, and then it can take between 7 to 12 years to utilize conventional methods to develop a new cultivar (Baenziger et al., 2006). The ability to fine-tune the expression of a quantitative trait locus rather than utilizing only what is available in wild relatives has already shown promise as a way to increase yield. For example, researchers edited genes in three pathways that contribute to productivity in tomato plants—plant architecture, fruit

BOX 2-2
Successful Engineering of Photosynthesis

Photosynthesis is the engine for life on Earth, yet the efficiency of solar energy conversion is relatively low. As such, it has high engineering potential (Schuler et al., 2016). There is a large international effort taking many different approaches to improve photosynthetic efficiency, including optimizing canopies, relaxing photoprotection, and bypassing photorespiration (Ort et al., 2015). Converting a C_3 plant into a C_4 plant that is much more efficient at converting sunlight is another effort under way to increase photosynthetic efficiency (Long, 2015; Ort et al., 2015). Using constitutive expression of the maize *GOLDEN2-LIKE* gene, researchers have now shown that they can confer a set of C_4-like traits on C_3 rice (Wang et al., 2017). These traits include the accumulation of photosynthetic enzymes and more interconnections between cells. Another exciting direction for increasing photosynthetic efficiency is to expand the spectral range for photosynthesis by allowing plants to absorb far-red light (Blakenship and Chen, 2013). The "first" Green Revolution improved yield potential but did so without improving solar energy conversion, making improvement of photosynthesis the basis for a second green revolution.

size, and inflorescence—to rapidly produce alleles that alter their promoters. The resultant plants displayed a series of previously unobserved phenotypes, including several with increased yield (Rodríguez-Leal et al., 2017).

Gene editing has now opened the door for quickly exploring additional changes. Box 2-2 describes a success story of a transgenic approach used in the quest to turn a C_3 plant (such as rice) into a C_4 plant (such as maize). Gene editing also offers the ability to eliminate linkage drag, a problem that has plagued crop breeding since its inception. Linkage drag occurs when undesirable traits are linked to, and therefore inherited with, desirable traits (Giovannoni, 2018; Zhu et al., 2018b). Using a variety of genomic and associated approaches to characterize a large tomato diversity population, Zhu and colleagues (2018b) were able to "see" the genomic outcomes of tomato domestication, many of which are undesirable. Knowledge of these "hitchhiking genes" is a necessary first step toward their eventual modification or elimination by gene-editing methodologies.

3.2 Development of Dynamic Crops

Agriculture could benefit from crops that provide farmers with the flexibility to change their crop's physiology in response to unexpected environments. Park and colleagues (2015) designed such a crop whose water use can be controlled by chemical application (see Box 2-3). An even greater

BOX 2-3
Synthetic Biology for Dynamic Crops

An ongoing focus of plant breeding is to create drought-tolerant plants, but the drought protective response in plants is often associated with "yield drag" (Yang et al., 2010). Recent work has demonstrated that it may be possible to alter a plant's use of water dynamically, essentially turning the demand for water on or off when needed. By genetically engineering the abscisic acid (ABA) protein receptor that naturally plays a role in controlling plant water use to bind to a synthetic agrochemical, it was possible to control the plant's drought protection response (Park et al., 2015). When tomato plants containing the engineered ABA were sprayed with the chemical, they used less water. The system is ready for real-world testing, suggesting that uses of synthetic biology of this type could be a near-term goal. Many aspects of plant physiology should be amenable to this approach, which could help usher in a more responsive agriculture.

advancement would involve developing crops that can respond in real time to environmental changes (e.g., drought, floods, and temperature extremes) as well as to the appearance of disease organisms and pests. Given the inability to reliably predict the conditions that any particular crop may face over the growing season combined with more extreme weather events predicted over the long term (Walthall et al., 2012), dynamic crops could better respond to such events and become an asset for enhancing food security.

3.3 Mining Plant Diversity

Traditionally, the traits to overcome the many challenges previously noted have come from wild relatives. Although gene editing can eliminate the laborious introgression of traits (at least for crops for which those traits can be successfully introduced; see discussion of transformation later in this chapter), there is a vast array of plant species whose agricultural potential remains untapped (see Box 2-4). There are more than 50,000 edible plants, but only 15 crops are used to meet 90 percent of the world's energy demands, and 3 commodity crops (rice, maize, and wheat) account for two-thirds of human caloric intake (Gruber, 2017). Additionally, more than 70 percent of wild relatives of domesticated crops are in urgent need of conservation, and are ironically threatened by the expansion of agriculture into natural ecosystems (Castañeda Alvarez et al., 2016). Around the world, gene banks and botanical gardens hold more than 7.4 million seeds or plant tissues from thousands of species (Gruber, 2017). These collections need to be maintained, curated, and explored.

BOX 2-4
Pichuberry: A Crop for Rapid Domestication

Cape gooseberry or pichuberry (*Physalis peruviana*) is native to the Andean region of South America. It is highly nutritious and can be eaten as fresh fruit or can be used to make juice or jam. It is gaining in popularity but it has a number of "wild" characteristics that prevent it from being easily cultivated. Knowledge of the genes related to the improvement and domestication of the tomato, a distant relative of *P. peruviana*, is motivating scientists to identify similar genes in the undomesticated pichuberry that could be targeted for domestication. Gene editing of *P. peruviana*'s genetic ortholog of the tomato gene *CLAVATA1* (*SlCLV1*), which controls meristem proliferation, gave rise to plants with narrow leaves and flowers with more organs (Van Eck et al., 2017). This offers proof of principle of rapid, targeted domestication using gene editing.

3.4 Taking Advantage of Plant–Microbe Interactions

Billions of microorganisms and macroorganisms (from viruses to nematodes) live on, inside, and near plants, both above and below ground (Leach et al., 2017). Beneficial plant–microbe interactions include direct stimulation of plant growth, protection of plants from pathogens and insect pests through direct production of toxins or through induced resistance in the plant, and improvement of resilience to environmental stress (e.g., drought, salinity) (Parnell et al., 2016; Finkel et al., 2017). Beneficial interactions can occur in the root zone (rhizosphere), on leaf surfaces (phyllosphere), and in internal tissues (endosphere) (APS, 2016).

The greatest density and diversity of microbial life associated with plants occurs in the soil, in the rhizosphere. In leguminous plants, some species of bacteria (*Rhizobium*) induce the formation of root nodules in a symbiotic relationship that converts atmospheric and largely inert N_2 into ammonia (NH_3) and other molecular precursors that the plant uses in the biosynthesis of nucleotides, coenzymes, and amino acids. In many more species of plants, fungal symbionts (called arbuscular mycorrhizal fungi) form hyphae that increase the ability of plant roots to access minerals (particularly phosphorus) and water (Leach et al., 2017). It has been proposed that an opportunity lies in domesticating or improving specific plant-associated microbial communities to use them as soil or seed inoculants to improve plant growth (Parnell et al., 2016).

The characterization of microbial communities in concert with plants, which often cannot be cultured outside the plant environment, has been enabled by high-throughput gene sequencing of community DNA (metagenomics) and RNA (metatranscriptomics) (Schlatter et al., 2015; Nesme et al., 2016). In addition, progress has been made at the intersection of miniaturized microfluidics and imaging that permits real-time monitoring of root interactions with microbial species (Massalha et al., 2017).

Based on knowledge gained from these technologies, a future opportunity is using synthetic biology tools to design plant–microbe associations that improve crop productivity. Such associations can be approached by using gene editing in plants, microbes, or both. For example, plant genes controlling nodule formation by nitrogen-fixing rhizobacteria might be expanded to non-legume crops to reduce the need for fertilizer application, and microbial consortia present in the root zone could be engineered to produce novel plant growth promoters or protectants (Ahkami et al., 2017).

3.5 Making Food More Nutritious

Relatively little breeding effort has gone into making staple crops better sources of vitamins and minerals. More than 2 billion people suffer from micronutrient deficiencies because their plant-based diets do not provide sufficient nutrition. Factors that drive the modern food system include consistency, predictability, low cost, and high-edible yield, but the nutritional value of food produced for direct human consumption has not been a priority of the market (Dwivedi et al., 2017). There have been concerted efforts to survey cultivars for natural variation in levels of select nutrients. For example, a recent study examined vitamin E levels in maize grain and determined that only two loci are responsible for most of the variation (Diepenbrock et al., 2017). This discovery suggests that targeting these loci in any number of crops may increase vitamin E content. There has also been some success using transgenic approaches to increase nutritional content. For example, biofortified rice that meets dietary targets for iron and zinc and has no yield penalty in the field is a major breakthrough in this area (Trijatmiko et al., 2016). There are also opportunities to increase levels of important phytonutrients that have health benefits, such as polyphenols and carotenoids (Martin and Li, 2017; Yu and Tian, 2017).

Unfortunately, breeding almost exclusively for increased yield has made some crops less nutritious. For example, the concentrations of iron, zinc, and selenium in wheat have dropped by 28, 25, and 18 percent, respectively, all in the period between 1920 and 2000 (Gruber, 2016). Such decreases are thought to be a dilution effect as improvements in the amount of fixed carbon have been made without a proportional increase in mineral content (Marles, 2017). This results in lower mineral concentrations when expressed on a dry weight basis. Furthermore, recent studies are projecting that levels of zinc and iron in grains and legumes will continue to decline as

CO_2 levels continue to rise (Myers et al., 2014; Zhu et al., 2018a). Because most of the world relies on plants as their dietary source of these micronutrients, such decreases are a cause for concern. A key scientific question is whether these declines in micronutrients can be reversed with changes in plant traits. Therefore, more attention needs to be paid to the unintended consequences of breeding exclusively for yield.

3.6 Optimizing Crop Production Systems

The gap between actual yield and yield potential can be accounted for by several sources of variance, including genetics (G), environment (E), management practices (M), and socioeconomic factors (S). Numerous enabling technologies can be implemented in crop production systems to achieve both resilience and sustainability and to enable crop yields to reach their full potential. These include the use of precision agriculture to manage water and fertilizer use, and the use of data science to integrate information from field- and plant-based sensors and weather prediction parameters (for a more complete discussion, see Chapter 5, section 3.1, "Leveraging Advances in Microelectronics, Sensing, and Modeling" and Chapter 6, "Water-Use Efficiency and Productivity"). Breakthroughs in genomics, nanotechnology, and robotics along with improvements in computational, statistical, and modeling capabilities will make it possible for scientists and producers to make well-informed, data-driven decisions. For example, as it will be discussed in Chapter 7, development of high-throughput automated phenotyping capabilities can speed the process of breeding via the use of artificial intelligence and machine learning. However, in order to successfully model, manage, and predict crop production in any given location, better information is also needed on how different cropping management systems (e.g., use of cover crops and crop rotation) influence soil properties such as water storage capacity and nutrient availability.

The emerging field of plant nanobiotechnology promises transformative solutions for nondestructive monitoring of plant signaling pathways and metabolism (Kwak et al., 2017). This can increase plant tolerance (e.g., drought, disease, and soil nutrient deficiencies [Elmer and White, 2016; Wu et al., 2017]), alter photosynthesis (Giraldo et al., 2014), and enable plants to communicate their biochemical status (Wong et al., 2017). These discoveries will lead to the creation of "smart" plants that are more resilient to climate-induced stresses. Better understanding of the fundamental biophysical processes controlling nanomaterial–plant interactions will enable delivery of nanomaterials to precise locations in plants where they are needed to be active. The continuous real-time monitoring of plant heat status and the ability to combine these nano-enabled technologies with wireless soil sensors and automated water and nutrient delivery systems can lead to more precise delivery of nutrients and water, leading to more efficient use of inputs and greater yields.

To ensure that all aspects of GEMS are addressed in a comprehensive manner, there is an opportunity to assemble teams (including geneticists, soil scientists, agronomists, plant pathologists, and entomologists) to work together to evaluate the response of different crop genotypes to environmental stresses and crop management practices (Hatfield and Walthall, 2015), and to factor in socioeconomic issues such as costs, access to small growers, and technology adoption. Such a systems approach can help to identify the optimum combination of G × M × S for the current and anticipated E.

3.7 Controlled Environment Agriculture (CEA)

Controlled environment agriculture[1] (CEA) offers systems-level opportunities to increase the sustainability of some crops (e.g., fruits, vegetables, and herbs) by providing resource-efficient farming systems with respect to water and nutrient use. CEA can close the loop on nitrogen and phosphorus use (Thomaier et al., 2015), reduce food miles (Weber and Matthews, 2008; Cleveland et al., 2011; Nicholson et al., 2015), and lower water exports from water-poor regions to water-rich regions. Locating food production in urban areas could also help to lower food waste by decreasing the amount of time from harvest to consumption. CEA can also address important food safety issues such as *Escherichia coli* outbreaks (CDC, 2018). CEA can provide year-long growing seasons and protection against pests and diseases. Genetic approaches described in this chapter can be used to provide traits to plants to make them suitable for growth in CEA. The importance of CEA for improving water-use efficiency is discussed in Chapter 6.

4. GAPS

There is growing excitement in the plant sciences and breeding communities that we are on the cusp of a second green revolution. While the first green revolution was facilitated by the introduction of genes from other varieties and wild relatives, this second revolution will be fueled by basic research discoveries with model organisms and analyses of massive datasets that will combine to identify genes and regulatory sequences for targeted editing. However, the following gaps in knowledge and capabili-

[1]Controlled environment agriculture includes indoor production systems that range from low-technology covers for field crops to highly automated greenhouses, vertical farms, and recirculating aquaculture systems (RASs). The latter use the latest advances in hydroponics or aquaponics to increase productivity and improve water- and nutrient-use efficiencies. For example, a RAS is coupled with hydroponic plant production by using waste derived from fish production as fertilizer for plant growth (Badiola et al., 2018; Palm et al., 2018).

ties will have to be addressed before these technologies can be applied to a diversity of crops.

1. ***Knowledge of the genetic basis of the phenotypes and of allelic variants.*** Although the functions of many plant genes are known from studies in the major model plants, such as *Arabidopsis*, there is a lack of basic knowledge of the genetic basis of many traits in crop species. It logically follows that understanding gene function is necessary for the use of gene editing to create desirable traits. However, a gene-editing approach can be used as a discovery tool to knock out genes in crop plants to better understand their functional roles. Additionally, because the precision of gene editing permits the modification of as little as a single base pair in the genome, it can be used to create allelic variants, essentially producing a library of diverse forms of the same gene, with which to explore function and phenotype.
2. ***Phenotyping: Connecting genomic variation with phenotypic impact.*** Identification of agriculturally desirable traits will require the design and construction of above- and belowground phenotyping facilities and the development of data science that can correlate phenotype to genotype from vast visual, physiological, and "omics" resources. There is a need to increase the speed to characterize phenotype, to use new technologies that include overhead imaging (e.g., photo from a drone) and underground imaging to detect morphologies and traits, including chemical exudates (Das et al., 2015; Shi et al., 2016).
3. ***Ability to improve transformation efficiency of more plant species.*** Gene editing requires an ability to introduce DNA into plant tissue (called genetic transformation) and, in most cases, regenerate the tissues into transformed plants. Unfortunately, despite substantial progress in sequencing, assembling, and annotating genomes from a vast array of plant species, significant bottlenecks exist in the successful genetic transformation of most crops. Research is needed to improve plant cell and tissue culture, identify better methods of introducing genetic material, and modulate the plant development pathway to improve the receptivity, stability, and regrowth of the transformed tissue (Altpeter et al., 2016). Studies of how plants regenerate themselves from cuttings or after injury, for example, have begun to provide insight into the genetic network that controls cellular proliferation and reprogramming. Such insights could lead to better methods to increase the chances of growing whole plants from transformed cells and tissues (Ikeuchi et al., 2018). A second green revolution powered by gene editing will not be pos-

sible unless facile transformation protocols are developed for use on any strain of any crop in any laboratory. Similarly, the inability to transform wild plant species will prevent the use of gene editing to accelerate their domestication.

4. ***Workforce, education, and training (and funding).*** Advances arising out of basic science and technology remain relatively new and underexploited for germplasm improvement for most crops. This is due in part to the relatively small community of plant scientists educated to use science and technology, and the limited public and private investments for plant trait discovery and for translating such knowledge to enable crop discoveries. Investment would be necessary to maintain a pipeline of investigators who can push the limits of scientific inquiry and train the next generation of advanced plant scientists (USDA, 2015).

5. RECOMMENDATIONS FOR NEXT STEPS

Despite steady decreases in the funding of both basic and applied crop science research, U.S. scientists have continued to innovate and make cutting-edge discoveries. However, without sufficient support, U.S. plant scientists are likely to fall behind their counterparts in other countries in fully realizing gene editing as both a basic discovery and crop improvement tool. Since 2009, plant science funding in China has quadrupled and is supported with infrastructure, eclipsing the U.S. investment. Notable advancements in all aspects of plant science have been forthcoming from the Chinese (Chong and Xu, 2014). Recognizing that food security and international competitiveness are critical components of national security, the following steps are recommended to secure U.S. leadership in crop improvement:

1. **Continue to genetically dissect and then introduce desirable traits (increased photosynthetic efficiency; drought and flood tolerance; temperature extremes tolerance; disease and pest resistance; and improved taste, aroma, and nutrition) and remove undesirable traits from crop plants through the use of both traditional genetic approaches and targeted gene editing.** The goal will be to
   a. expand the number of alleles of known breeding improvement genes from what have been introgressed from wild relatives;
   b. engineer below- and aboveground plant architecture, by obtaining and applying basic knowledge of plant root, shoot, and influorescence development;
   c. modify the plant microbiome to enhance desirable crop traits, including resistance to disease and increased nutrient-use efficiency; and

   d. remove undesirable traits (due to negative epistasis) that are tightly linked to alleles selected during domestication.
2. **Enable routine genome modification of all crop plants through the development of facile transformation and regeneration technologies.** Achieving this goal will require research into improved plant cell and tissue culture, better methods of introducing genetic material, and the ability to modulate plant development to improve the regeneration of the transformed tissue into whole plants. It will also require development of facilities for more rapid phenotype detection and analysis under different environmental (soil, climate, and moisture) conditions and management regimes.
3. **Monitor plant stress and nutrients through the development of novel sensing technologies, and allow plants to better respond to environmental challenges (heat, drought, flood, pests, and nutrient requirements) by exploring the use of nanotechnology, synthetic biology, and the plant microbiome to develop dynamic crops that can turn certain functions on or off only when needed.** Developing dynamic crops will require complementary approaches to gene editing, including harnessing beneficial plant–root–soil–microbe interactions to enhance desirable crop traits (such as drought resistance, disease resistance, and nutrient-use efficiency), using novel sensing technologies to sense plant nutrient status and stress, and using nanotechnologies for delivering nutrients and managing plant stress.

Using the genetic diversity of plant and microbial life available together with new molecular and other tools is key for unlocking many opportunities for crop improvement. The process of prioritizing the most important opportunity to pursue would be best informed by thoughtful analyses of the targeted crops and their prospective new traits from a systems perspective, as described in Chapter 9. This includes envisioning their performance and impact in the context of the agroecological system interfacing with humans and socioeconomic considerations that play a role.

REFERENCES

Ahkami, A. H., R. A. White III, P. P. Handakumbura, and C. Jansson. 2017. Rhizosphere engineering: Enhancing sustainable plant ecosystem productivity. *Rhizosphere* 3(2):233-243.

Alston, J. M., J. M. Beddow, and P. G. Pardey. 2009. Agricultural research, productivity, and food prices in the long run. *Science* 325(5945):1209-1210.

Altpeter, F., N. M. Springer, L. E. Bartley, A. E. Blechl, T. P. Brutnell, V. Citovsky, L. J. Conrad, S. B. Gelvin, D. P. Jackson, A. P. Kausch, P. G. Lemaux, J. I. Medford, M. L. Orozco-Cárdenas, D. M. Tricoli, J. Van Eck, D. F. Voytas, V. Walbot, K. Wang, Z. J. Zhang, and C. N. Stewart. 2016. Advancing crop transformation in the era of genome editing. *Plant Cell* 28(7):1510-1520.

Andersen, M. A. 2015. Public investment in U.S. agricultural R&D and the economic benefits. *Food Policy* 51:38-43. Available at https://doi.org/10.1016/j.foodpol.2014.12.005 (accessed May 8, 2018).

Andersen, M. A., J. M. Alston, P. G. Pardey, and A. Smith. 2018. A century of U.S. farm productivity growth: A surge then a slowdown. *American Journal of Agricultural Economics* 100(4):1072-1090. Available at https://doi.org/10.1093/ajae/aay023 (accessed May 31, 2018).

APS (American Phytopathological Society). 2016. *Phytobiomes: A Roadmap for Research and Translation*. St. Paul, MN: American Phytopathological Society. Available at https://www.apsnet.org/members/outreach/ppb/Documents/PhytobiomesRoadmap.pdf (accessed May 8, 2018).

Badiola, M., O. Basurko, R. Piedrahita, P. Hundley, and D. Mendiola. 2018. Energy use in recirculating aquaculture systems (RAS): A review. *Aquacultural Engineering* 81. doi: 10.1016/j.aquaeng.2018.03.003.

Baenziger, P. S., W. K. Russell, G. L. Graef, and B. T. Campbell. 2006. Improving lives: 50 years of crop breeding, genetics and cytology (C-1). *Crop Science* 46:2230-2244.

Bailey-Serres, J., S. C. Lee, and E. Brinton. 2012. Waterproofing crops: Effective flooding survival strategies. *Plant Physiology* 160(4):1698-1709.

Baltes, N. J., J. Gil-Humanes, and D. F. Voytas. 2017. Chapter one–genome engineering and agriculture: Opportunities and challenges. *Progress in Molecular Biology and Translational Science* 149:1-26.

Baumhardt, R. L., B. A. Stewart, and U. M. Sainju, 2015. North American soil degradation: Processes, practices, and mitigating strategies: A review. *Sustainability* 7(3):2936-2960.

Blakenship, R. E., and M. Chen. 2013. Spectral expansion and antenna reduction can enhance photosynthesis for energy production. *Current Opinion in Chemical Biology* 17:457-461.

Castañeda Álvarez, N., C. K. Khoury, H. A. Achicanoy, V. Bernau, H. Dempewolf, R. J. Eastwood, L. Guarino, R. H. Harker, A. Jarvis, N. Maxted, J. V. Müller, J. Ramirez-Villegas, C. C. Sosa, P. C. Struik, H. Vincent, and J. Toll. 2016. Global conservation priorities for crop wild relatives. *Nature Plants* 2:16022. doi: 10.1038/nplants.2016.22.

CDC (Centers for Disease Control and Prevention). 2018. *Multistate Outbreak of E. coli O157:H7 Infections Linked to Romaine Lettuce, Investigation Notice: Multistate Outbreak of E. coli O157:H7 Infections April 2018*. Available at https://www.cdc.gov/ecoli/2018/o157h7-04-18/index.html (accessed June 4, 2018).

Chong, K. and Z. Xu. 2014. Investment in plant research and development bears fruit in China. *Plant Cell Reports* 33(4):541-550. doi: 10.1007/s00299-014-1587-6.

Cleveland, D. A., C. N. Radka, N. M. Müller, T. D. Watson, N. J. Rekstein, H. V. Wright, and S. E. Hollingshead. 2011. Effect of localizing fruit and vegetable consumption on greenhouse gas emissions and nutrition, Santa Barbara County. *Environmental Science & Technology* 45(10):4555-4562.

Court, C. D., A. W. Hodges, M. Ramani, and T. H. Spleen. 2017. *Economic Contributions of the Florida Citrus Industry: 2015-2016*. Gainsville: University of Florida Food and Economics Department.

Das, A., H. Schneider, J. Burridge, A. K. M. Ascanio, T. Wojciechowski, C. N. Topp, J. P. Lynch, J. S. Weitz, and A. Bucksch. 2015. Digital imaging of root traits (DIRT): A high-throughput computing and collaboration platform for field-based root phenomics. *Plant Methods* 11:51. Available at http://doi.org/10.1186/s13007-015-0093-3 (accessed June 4, 2018).

Diepenbrock, C. H., C. B. Kandianis, A. E. Lipka, M. Magallanes-Lundback, B. Vaillancourt, E. Góngora-Castillo, J. G. Wallace, J. Cepela, A. Mesberg, P. J. Bradbury, D. C. Ilut, M. Mateos-Hernandez, J. Hamilton, B. F. Owens, T. Tiede, E. S. Buckler, T. Rocheford, C. R. Buell, M. A. Gore, and D. DellaPenna. 2017. Novel loci underlie natural variation in vitamin E levels in maize grain. *Plant Cell* 29(10):2374-2392.

Diffenbaugh, N. S., D. L. Swain, and D. Touma. 2015. Anthropogenic warming has increased drought risk in California. *Proceedings of the National Academy of Sciences of the United States of America* 112(13):3931-3936.

Dwivedi, S. L., E. T. Lammerts van Bueren, S. Ceccarelli, S. Grando, H. D. Upadhyaya, and R. Ortiz. 2017. Diversifying food systems in the pursuit of sustainable food production and healthy diets. *Trends in Plant Science* 22(10):842-856.

Elmer, W. H., and J. C. White. 2016. The use of metallic oxide nanoparticles to enhance growth of tomatoes and eggplants in disease infested soil or soilless medium. *Environmental Science: Nano* 3:1072-1079.

Evans, L. T., and R. A. Fischer. 1999. Yield potential: Its definition, measurement, and significance. *Crop Science* 39(6):1544-1551.

Fernandez-Cornejo, J. 2004. *The Seed Industry in U.S. Agriculture: An Exploration of Data and Information on Crop Seed Markets, Regulation, Industry Structure, and Research and Development.* Washington, DC: USDA Economic Research Service. Available at https://www.ers.usda.gov/publications/pub-details/?pubid=42531 (accessed May 10, 2018).

Finkel, O. M., G. Castrillo, S. Herrera Paredes, I. Salas González, and J. L. Dangl. 2017. Understanding and exploiting plant beneficial microbes. *Current Opinion in Plant Biology* 38:155-163.

Giovannoni, J. 2018. Tomato multiomics reveals consequences of crop domestication and improvement. *Cell* 172(1):6-8.

Giraldo, J. P., M. P. Landry, S. M. Faltermeier, T. P. McNicholas, N. M. Iverson, A. A. Boghossian, N. F. Reuel, A. J. Hilmer, F. Sen, J. A. Brew, and M. S. Strano. 2014. Plant nanobionics approach to augment photosynthesis and biochemical sensing. *Nature Materials* 13:400-408.

Gruber, K. 2016. Re-igniting the green revolution with wild crops. *Nature Plants* 2:16048.

Gruber, K. 2017. Agrobiodiversity: The living library. *Nature* 544(7651):S8-S10.

Hatfield, J. L., and C. L. Walthall. 2015. Meeting global food needs: Realizing the potential via genetics × environment × management interactions. *Agronomy Journal* 107:1215.

Howitt, R., D. MacEwan, J. Medellin-Azuara, J. Lund, and D. Sumner. 2015. *Economic Analysis of the 2015 Drought for California Agriculture.* Available at https://watershed.ucdavis.edu/files/biblio/Final_Drought%20Report_08182015_Full_Report_WithAppendices.pdf (accessed May 8, 2018).

Ikeuchi, M., M. Shibata, B. Rymen, A. Iwase, A. M. Bågman, L. Watt, D. Coleman, D. S. Favero, T. Takahashi, S. E. Ahnert, S. M. Brady, and K. Sugimoto. 2018. A gene regulatory network for cellular reprogramming in plant regeneration. *Plant Cell Physiology* 59(4):765-777. doi: 10.1093/pcp/pcy013.

Katsaruware-Chapoto, R. D., P. L. Mafongoya, and A. Gubba. 2017. Responses of insect pests and plant diseases to changing and variable climate: A review. *Journal of Agricultural Science* 9(12):160.

Konikow, L. F. 2013. *Groundwater Depletion in the United States (1900-2008)*. Scientific Investigations Report 2013-5079. Reston, VA: U.S. Geological Survey. Available at http://pubs.usgs.gov/sir/2013/5079 (accessed May 10, 2018).

Kwak, S.-Y., M. H. Wong, T. T. S. Lew, G. Bisker, M. A. Lee, A. Kaplan, J. Dong, A. T. Liu, V. B. Koman, R. Sinclair, C. Hamann, and M. S. Strano. 2017. Nanosensor technology applied to living plant systems. *Annual Review of Analytical Chemistry* 10:113-140.

Leach, J. E., L. R. Triplett, C. T. Argueso, and P. Trivedi. 2017. Communication in the phytobiome. *Cell* 169(4):587-596.

Livingston, M., J. Fernandez-Cornejo, J. Unger, C. Osteen, D. Schimmelpfennig, T. Park, and D. Lambert. 2015. *The Economics of Glyphosate Resistance Management in Corn and Soybean Production*. Washington, DC: U.S. Department of Agriculture's Economic Research Service.

Lobell, D. B., and C. B. Field. 2007. Global scale climate–crop yield relationships and the impacts of recent warming. *Environmental Research Letters* 2(1):014002.

Long, S. 2015. Meeting the global food demand of the future by engineering crop photosynthesis and yield potential. *Cell* 161(1):56-66.

Mallakpour, I., and G. Villarini. 2015. The changing nature of flooding across the central United States. *Nature Climate Change* 5(3):250.

Marles, R. 2017. Mineral nutrient composition of vegetables, fruits and grains: The context of reports of apparent historical declines. *Journal of Food Composition Analysis* 56:93-103.

Martin, C., and J. Li. 2017. Medicine is not health care, food is health care: Plant metabolic engineering, diet and human health. *New Phytologist* 216(3):699-719.

Massalha, H., E. Korenblum, S. Malitsky, O. H. Shapiro, and A. Aharoni. 2017. Live imaging of root–bacteria interactions in a microfluidics setup. *Proceedings of the National Academy of Sciences of the United States of America* 114(17):4549-4554.

Myers, S. S., A. Zanobetti, I. Kloog, P. Huybers, A. D. B. Leakey, A. J. Bloom, E. Carlisle, L. H. Dietterich, G. Fitzgerald, T. Hasegawa, N. M. Holbrook, R. L. Nelson, M. J. Ottman, V. Raboy, H. Sakai, K. A. Sartor, J. Schwartz, S. Seneweera, M. Tausz, and Y. Usui. 2014. Increasing CO_2 threatens human nutrition. *Nature* 510(7503):139.

NASEM (National Academies of Sciences, Engineering, and Medicine). 2018. *A Review of the Citrus Greening Research and Development Efforts Supported by the Citrus Research and Development Foundation: Fighting a Ravaging Disease*. Washington, DC: The National Academies Press.

Nesme, J., W. Achouak, S. N. Agathos, M. Bailey, P. Baldrian, D. Brunel, Å. Frostegård, T. Heulin, J. K. Jansson, E. Jurkevitch, K. L. Kruus, G. A. Kowalchuk, A. Lagares, H. M. Lappin-Scott, P. Lemanceau, D. Le Paslier, I. Mandic-Mulec, J. C. Murrell, D. D. Myrold, R. Nalin, P. Nannipieri, J. D. Neufeld, F. O'Gara, J. J. Parnell, A. Pühler, V. Pylro, J. L. Ramos, L. F. W. Roesch, M. Schloter, C. Schleper, A. Sczyrba, A. Sessitsch, S. Sjöling, J. Sørensen, S. J. Sørensen, C. C. Tebbe, E. Topp, G. Tsiamis, J. D. van Elsas, G. van Keulen, F. Widmer, M. Wagner, T. Zhang, X. Zhang, L. Zhao, Y.-G. Zhu, T. M. Vogel, and P. Simonet. 2016. Back to the future of soil metagenomics. *Frontiers in Microbiology* 7:73.

Nicholson, C. F., X. He, M. I. Gómez, H. O. Gao, and E. Hill. 2015. Environmental and economic impacts of localizing food systems: The case of dairy supply chains in the northeastern United States. *Environmental Science & Technology* 49(20):12005-12014.

Nielsen, R. L. 2017. *Historical Corn Grain Yields for the U.S.* Web publication, Purdue University Agricultural Extension. Available at https://www.agry.purdue.edu/ext/corn/news/timeless/yieldtrends.html (accessed June 13, 2018).

Ort, D. R., S. S. Merchant, J. Alric, A. Barkan, R. E. Blankenship, R. Bock, R. Croce, M. R. Hanson, J. M. Hibberd, S. P. Long, T. A. Moore, J. Moroney, K. K. Niyogi, M. A. J. Parry, P. P. Peralta-Yahya, R. C. Prince, K. E. Redding, M. H. Spalding, K. J. van Wijk, W. F. J. Vermaas, S. von Caemmerer, A. P. M. Weber, T. O. Yeates, J. S. Yuan, and X. G. Zhu. 2015. Redesigning photosynthesis to sustainably meet global food and bioenergy demand. *Proceedings of the National Academy of Sciences of the United States of America* 112(28):8529-8536.

Palm, H. W., U. Knaus, S. Appelbaum, S. Goddek, S. M. Strauch, T. Vermeulen, M. H. Jijakli, and B. Kotzen. 2018. Towards commercial aquaponics: A review of systems, designs, scales and nomenclature. *Aquaculture International* 26:813-842.

Pardey, P. G., and J. M. Beddow. 2013. *Agricultural Innovation: The United States in a Changing Global Reality.* Chicago: Chicago Council on Global Affairs. Available at https://www.thechicagocouncil.org/sites/default/files/Agricultural_Innovation_Final%281%29.pdf (accessed May 8, 2018).

Park, S., F. C. Peterson, A. Mosquna, J. Yao, B. F. Volkman, and S. R. Cutler. 2015. Agrochemical control of plant water use using engineered abscisic acid receptors. *Nature* 520(7548):545.

Parnell, J. J., R. Berka, H. A. Young, J. M. Sturino, Y. Kang, D. M. Barnhart, and M. V. DiLeo. 2016. From the lab to the farm: An industrial perspective of plant beneficial microorganisms. *Frontiers in Plant Science* 7:1110.

Ray, D. K., N. Ramankutty, N. D. Mueller, P. C. West, and J. A. Foley. 2012. Recent patterns of crop yield growth and stagnation. *Nature Communications* 3:1293.

Rockström, J., J. Williams, G. Daily, A. Noble, N. Matthews, L. Gordon, H. Wetterstrand, F. DeClerck, M. Shah, P. Steduto, C. de Fraiture, N. Hatibu, O. Unver, J. Bird, L. Sibanda, and J. Smith 2017. Sustainable intensification of agriculture for human prosperity and global sustainability. *AMBIO* 46(1):4-17.

Rodríguez-Leal, D., Z. H. Lemmon, J. Man, M. E. Bartlett, and Z. B. Lippman. 2017. Engineering quantitative trait variation for crop improvement by genome editing. *Cell* 171(2):470-480.

Savary, S., A. Ficke, J. Aubertot, and C. Hollier. 2012. Crop losses due to diseases and their implications for global food production losses and food security. *Food Security* 4(4):519-537.

Schlatter, D. C., M. G. Bakker, J. M. Bradeen, and L. L. Kinkel. 2015. Plant community richness and microbial interactions structure bacterial communities in soil. *Ecology* 96(1):134-142.

Schuler, M. L., O. Mantegazza, and A. P. M. Weber. 2016. Engineering C_4 photosynthesis into C_3 chassis in the synthetic biology age. *Plant Journal* 87(1):51-65.

Shi, Y., J. A. Thomasson, S. C. Murray, N. A. Pugh, W. L. Rooney, S. Shafian, N. Rajan. G. Rouze, C. L. Morgan, H. L. Neely, A. Rana, M. V. Bagavathiannan, J. Henrickson, E. Bowden, J. Valasek, J. Olsenholler, M. P. Bishop, R. Sheridan, E. B. Putman, S. Popescu, T. Burks, D. Cope, A. Ibrahim, B. F. McCutchen, D. D. Baltensperger, R. V. Avant, M. Vidrine, and C. Yang. 2016. Unmanned aerial vehicles for high-throughput phenotyping and agronomic research. *PLoS ONE* 11(7):e0159781.

Steward, D. R., P. J. Bruss, X. Yang, S. A. Staggenborg, S. M. Welch, and M. D. Apley. 2013. Tapping unsustainable groundwater stores for agricultural production in the High Plains Aquifer of Kansas, projections to 2110. *Proceedings of the National Academy of Sciences of the United States of America* 110(37):E3477-E3486.

Thomaier, S., K. Specht, D. Henckel, and A. Dierich. 2015. Farming in and on urban buildings: Present practice and specific novelties of Zero-Acreage Farming (ZFarming). *Renewable Agriculture and Food Systems* 30:43-54.

Trijatmiko, K. R., C. Dueñas, N. Tsakirpaloglou, L. Torrizo, F. M. Arines, C. Adeva, J. Balindong, N. Oliva, M. V. Sapasap, J. Borrero, J. Rey, P. Francisco, A. Nelson, H. Nakanishi, E. Lombi, E. Tako, R. P. Glahn, J. Stangoulis, P. Chadha-Mohanty, A. A. T. Johnson, J. Tohme, G. Barry, and I. H. Slamet-Loedin. 2016. Biofortified indica rice attains iron and zinc nutrition dietary targets in the field. *Scientific Reports* 6:19792.

UN DESA (United Nations Department of Economic and Social Affairs). 2017. *World Population Prospects: The 2017 Revision, Key Findings and Advance Tables.* ESA/P/WP/248. Available at https://esa.un.org/unpd/wpp/publications/Files/WPP2017_KeyFindings.pdf (accessed May 10, 2018).

USDA (U.S. Department of Agriculture). 2015. *USDA Roadmap for Plant Breeding.* Available at https://www.usda.gov/sites/default/files/documents/usda-roadmap-plant-breeding.pdf (accessed May 8, 2018).

USDA-ERS (U.S. Department of Agriculture's Economic Research Service). 2018. *Farm Income and Wealth Statistics: Cash Receipts by State.* Available at https://data.ers.usda.gov/reports.aspx?ID=17843 (accessed June 18, 2018).

USDA-FAS (U.S. Department of Agriculture's Foreign Agricultural Service). 2018a. *Global Agricultural Trade System (GATS) Data.* Available at https://apps.fas.usda.gov/gats/detectscreen.aspx?returnpage=default.aspx (accessed May 8, 2018).

USDA-FAS. 2018b. *World Agricultural Production.* Available at https://apps.fas.usda.gov/psdonline/circulars/production.pdf (accessed May 8, 2018).

USDA-RMA (U.S. Department of Agriculture's Risk Management Agency). 2018. Cause of loss historical data files. Available at https://www.rma.usda.gov/SummaryOfBusiness/CauseOfLoss (accessed July 11, 2018).

USTR (Office of the U.S. Trade Representative). 2017. *USTR Success Stories: Opening Markets for U.S. Agricultural Exports.* Available at https://ustr.gov/about-us/policy-offices/press-office/fact-sheets/2017/march/ustr-success-stories-opening-markets-us (accessed May 8, 2018).

Van Eck, J., K. Swartwood, Z. Lemmone, J. Dalrymple, and Z. B. Lippman. 2017. Development of *Agrobacterium*-mediated transformation of *Physalis peruviana* and application of CRISPR/Cas9 genome editing. *In Vitro Cellular & Developmental Biology—Animal* 53(Suppl.):S34-S35.

Walthall, C. L., J. Hatfield, P. Backlund, L. Lengnick, E. Marshall, M. Walsh, S. Adkins, M. Aillery, E. A. Ainsworth, C. Ammann, C. J. Anderson, I. Bartomeus, L. H. Baumgard, F. Booker, B. Bradley, D. M. Blumenthal, J. Bunce, K. Burkey, S. M. Dabney, J. A. Delgado, J. Dukes, A. Funk, K. Garrett, M. Glenn, D. A. Grantz, D. Goodrich, S. Hu, R. C. Izaurralde, R. A. C. Jones, S.-H. Kim, A. D. B. Leaky, K. Lewers, T. L. Mader, A. McClung, J. Morgan, D. J. Muth, M. Nearing, D. M. Oosterhuis, D. Ort, C. Parmesan, W. T. Pettigrew, W. Polley, R. Rader, C. Rice, M. Rivington, E. Rosskopf, W. A. Salas, L. E. Sollenberger, R. Srygley, C. Stöckle, E. S. Takle, D. Timlin, J. W. White, R. Winfree, L. Wright-Morton, and L. H. Ziska. 2012. *Climate Change and Agriculture in the United States: Effects and Adaptation.* USDA Technical Bulletin 1935. Washington, DC: U.S. Department of Agriculture's Agricultural Research Service.

Wang, P., R. Khoshravesh, S. Karki, R. Tapia, C. P. Balahadia, A. Bandyopadhyay, W. Paul Quick, R. Furbank, T. L. Sage, and J. A. Langdale. 2017. Recreation of a key step in the evolutionary switch from C_3 to C_4 leaf anatomy. *Current Biology* 27(21):3278-3287.

Weber, C. L., and H. S. Matthews. 2008. Food-miles and the relative climate impacts of food choices in the United States. *Environmental Science & Technology* 42(10):3508-3513.

Wei, X., Z. Zhang, P. Shi, P. Wang, Y. Chen, X. Song, and F. Tao. 2015. Is yield increase sufficient to achieve food security in China? *PLoS ONE* 10(2):e0116430.

Wong, M. H., J. P. Giraldo, S.-Y. Kwak, V. B. Koman, R. Sinclair, T. T. Salim Lew, G. Bisker, P. Liu, and M. S. Strano. 2017. Nitroaromatic detection and infrared communication from wild-type plants using plant nanobionics. *Nature Materials* 16(2):264-272.

Wu, H., N. Tito, and J. P. Giraldo. 2017. Anionic cerium oxide nanoparticles protect plant photosynthesis from abiotic stress by scavenging reactive oxygen species. *ACS Nano* 11(11):11283-11297.

Yang, S., B. Vanderbeld, J. Wan, and Y. Huang. 2010. Narrowing down the targets: Towards successful genetic engineering of drought-tolerant crops. *Molecular Plant* 3(3):469-490.

Yin, X., A. K. Biswal, J. Dionora, K. M. Perdigon, C. P. Balahadia, S. Mazumdar, C. Chater, H.-C. Lin, R. A. Coe, T. Kretzschmar, J. E. Gray, P. W. Quick, and A. Bandyopadhyay. 2017. CRISPR-Cas9 and CRISPR-Cpf1 mediated targeting of a stomatal developmental gene EPFL9 in rice. *Plant Cell Reports* 36(5):745-757.

Yu, S., and L. Tian. 2017. Breeding major cereal grains through the lens of nutrition-sensitivity. *Molecular Plant* 11:23-30.

Zhu, C., K. Kobayashi, I. Loladze, J. Zhu, Q. Jiang, X. Yu, G. Liu, S. Seneweera, K. L. Ebi, A. Drewnowski, N. K. Fukagawa, and L. H. Ziska. 2018a. Carbon dioxide (CO_2) levels this century will alter the protein, micronutrients, and vitamin content of rice grins with potential health consequences for the poorest rice-dependent countries. *Science Advances* 4(5):eaaq1012.

Zhu, G., S. Wang, Z. Huang, S. Zhang, Q. Liao, C. Zhang, T. Lin, M. Qin, M. Peng, C. Yang, X. Cao, X. Han, X. Wang, E. van der Knaap, Z. Zhang, X. Cui, H. Klee, A. R. Fernie, and J. Luo. 2018b. Rewiring of the fruit metabolome in tomato breeding. *Cell* 172(1):249-261.

3

Animal Agriculture

1. INTRODUCTION

Animal products are a primary source of protein and key nutrients in American diets (Bentley, 2017). In addition, livestock and poultry production account for approximately $100 billion per year in agricultural cash receipts (USDA-ERS, 2018a). In the United States, most food animal production (meat, fish, milk, and eggs) is accomplished through an intensive rearing system that reflects decades of improvements in production efficiencies made possible by research and development. Genetic improvement and adoption of optimized nutritional programs, along with innovations to maintain and improve animal health status, have reduced the costs of production, lowered prices for consumers, decreased resources used (resulting in lower greenhouse gas [GHG] emission intensities per unit of production), and increased the competitiveness of American products internationally, benefiting both local and national economies (Havenstein et al., 2003; Capper et al., 2009; Capper, 2011; Gerber et al., 2011; Tokach et al., 2016). For example, the GHG emissions associated with the production of a glass of milk in the United States in 1977 were one-third of what they were in 1944 (Capper et al., 2009); today, livestock sources (including enteric fermentation and manure) account for about 3.9 percent of U.S. anthropogenic GHG emissions expressed as carbon dioxide equivalents (EPA, 2018). Over the past 15 years, the U.S. livestock industry has gained greater access to international markets with a growing share of its production destined for foreign markets. In 2016, exports accounted for 23 percent of pork produced, nearly 20 percent for broiler meat, 14 percent for

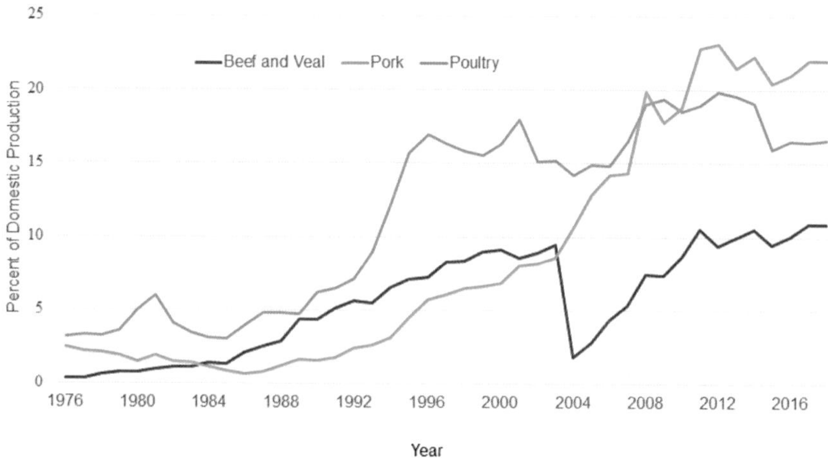

FIGURE 3-1 U.S. meat exports (beef and veal, pork, and poultry) as a share of domestic production from 1976 to 2016.
NOTE: The unit description for beef and veal and for pork is 1,000 metric tons per carcass-weight equivalent, while the unit description for poultry is 1,000 metric tons.
SOURCE: USDA-FAS, 2018.

turkey, and 11 percent for beef (see Figure 3-1). This brings in more than $25 billion annually in export sales (USDA-ERS, 2018b) to the national economy.

2. CHALLENGES

The science that produced advances in animal agricultural productivity and efficiency in the past will need to be accelerated and expanded in scope in the decade ahead. Recent analyses have warned that the agricultural production worldwide will have collective difficulty in meeting global demand for food in 2050 as a consequence of the growing world population, with demands increasing in the range of 59 to 98 percent (Ingram et al., 2010; Valin et al., 2014; UN DESA, 2015). The production required to meet that demand will occur within the United States and across a global landscape, including places with different environments, production systems, and animal breeds. As a world leader in efficient animal production, it is in the best interests of the United States to continue research in approaches to efficiently address the world's demand for animal products.

There are several compelling reasons for investing in animal agricultural research to meet this challenge. First, the increasing global demand for animal-source food will impact domestic prices, and therefore continuing to augment efficiencies is essential to keep food affordable for U.S. consumers. Although it is beyond the scope of this report to comment on the American landscape of chronic diseases related to dietary choices, hunger and hidden hunger remain problematic in the developing world. Cognitive stunting due to lack of micronutrients, which are bioavailable predominantly in animal-source foods, is a major problem that results in whole generations of children who cannot reach their cerebral potential and consequently a corresponding drop in country gross domestic product potential (Galasso and Wagstaff, 2018). Despite Western-centric concerns that overconsumption of meat is detrimental to health in the developed world, the situation is the reverse for the world's poorest as the underconsumption of animal-source food is detrimental to health (James and Palmer, 2015; Beal et al., 2017). It is this latter population that, once given the economic liberty to choose their foodstuffs, will select more animal-source foods and create a larger global demand. Second, with increasing efficiencies and intensity of production, animals need to be reared in systems that promote animal welfare, minimize GHG emissions and pollution, and decrease the potential of foodborne illness. Third, animal diseases continue to move throughout the world which at a minimum decrease productivity, while transboundary animal diseases (which are high-consequence contagious animal diseases) can bankrupt animal industries in the United States and globally.

2.1 Increased Demand for Animal Protein

Population growth combined with rising incomes in the developing world will result in the need to increase animal production to meet domestic and export market demands. Animal protein production has made impressive advances in the past 50 years, but more will be required to meet the projected demands of a growing global population (Ingram et al., 2010; UN DESA, 2015). On a per capita basis, consumption of animal-source foods (defined as meat, eggs, dairy, and fish) has steadily increased in the United States and is projected to continue to increase globally (NOAA, 2015; USDA-ERS, 2018a). As low-income countries become emerging economies, consumption of animal-source foods will increase, and so the "bottom billion" will begin to get a share of nutrients that promotes better health by supplying adequate protein and nutrients through animal-source foods. Figure 3-2 shows the levels of animal products produced globally from 1980 to the present, with numbers projected to 2030. Predictions are that demand for animal-derived food in 2050 could be 70 percent higher than 2005 levels, with the demand for beef and pork increasing by as much as

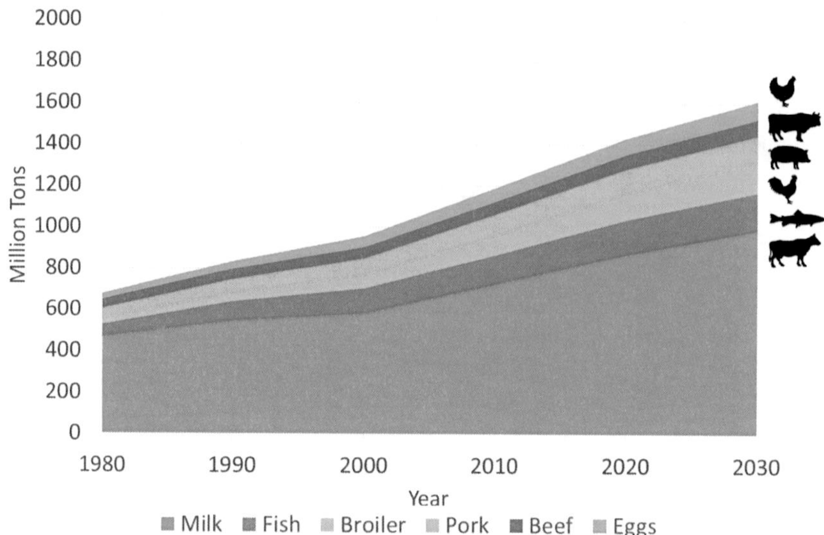

FIGURE 3-2 Global milk, fish, broiler, pork, beef, and egg production since 1980 and projected to 2030.
SOURCE: Alexandratos and Bruinsma, 2012.

66 and 43 percent, respectively (Alexandratos and Bruinsma, 2012). The highest growth is expected for poultry meat at 121 percent growth, especially in developing countries, with demand for eggs potentially increasing by 65 percent (Mottet and Tempio, 2017). As a result, livestock numbers are expected to continue to increase significantly, although at lower rates than in past years (Alexandratos and Bruinsma, 2012).

With respect to the contribution of fish to 2030 animal protein demands, the capture fisheries supply is unlikely to increase as stocks of key species are already fully harvested, and ocean acidification from cumulative GHG emissions over several decades is leading to biodiversity loss and threatening single-cell phytoplankton—the base of marine food chains that account for more than 50 percent of the photosynthesis and oxygen production on earth. Eutrophication and algal blooms resulting from the widespread use and runoff of nitrogen and phosphorous-based fertilizers into lakes, rivers, and coastal estuaries are further depleting aquatic plant and animal species, many of which provide food for humans (Carpenter et al., 1998; Canfield et al., 2010).

Aquaculture is therefore expected to dominate growth in the fish sector. Fishery production is expected to expand by more than 30 metric tons by 2030, 95 percent of which will come from developing countries. China is likely to have a significant influence on global fish markets, accounting for 37 percent of total fish production in 2030. Assumptions suggest that steady improvements in feed and feeding efficiency within the aquaculture sector will contribute to its growth (World Bank, 2013), and recirculating aquaculture systems could contribute to more efficient water and nutrient use (see Chapter 6 for further discussion on Controlled Environments). Food security in future years will surely depend on increasing the food-provisioning capacity of aquatic as well as terrestrial ecosystems.

2.2 Animal Diseases

Animal diseases are responsible for an average loss of more than 20 percent of animal production worldwide and can have major economic impacts (OIE, 2018). Export partners can justifiably block imports and issue trade sanctions if zoonotic or transboundary animal diseases, such as highly pathogenic avian influenza, occur in the United States. Blocked exports can cause a glut of the commodity for domestic consumption, resulting in ripple effects in the economy, from decreased prices for consumers to very negative farm-level impacts in the rural economy. Introduction of foot-and-mouth disease into the United States remains a constant threat, with estimates of the economic burden ranging from $37 billion to $228 billion (Oladosu et al., 2013). Box 3-1 provides two examples in which

BOX 3-1
Recent Cases of Animal Diseases and Their Economic Impact

Highly pathogenic avian influenza entered the United States through migratory birds and caused an extensive outbreak in the Midwest that extended through mid-2015. The U.S. Department of Agriculture (USDA) undertook the most aggressive foreign animal disease control efforts ever by depopulating 7.5 million turkeys and 42 million chickens in an effort that cost U.S. taxpayers $879 million. The decision to eradicate was made because of the costs associated with not controlling the disease, which, through simulations, would have averaged $1.53 billion.

Bovine spongiform encephalopathy was found in late 2003 in one cow for the first time in the United States. This single case of "mad cow disease" prompted massive closure of major U.S. beef export markets, with 80 percent of exports blocked for the following year, costing the industry between $3.2 billion and $4.7 billion.

SOURCES: Coffey et al., 2005; USDA-APHIS, 2016.

recent cases of high-consequence animal diseases had major economic impacts. Even endemic diseases, such as bovine respiratory disease—the leading cause of morbidity and mortality in cattle that is caused by a variety of viruses and bacteria—can greatly hamper an animal's productivity, may result in death, and is estimated to cost the cattle industry approximately $3 billion annually (Griffin, 1997).

2.3 Sustainability in Animal Agriculture

There is increased public interest in environmental, economic, and animal welfare sustainability questions related to animal agriculture. Unfortunately, there are no clearly defined objective measures for all of the sustainability questions related to animal agriculture, which in itself represents a notable area for investigation. Sustainability goals are sometimes in conflict, and managing a system for optimal environmental stewardship may clash with economic or animal welfare objectives (Llonch et al., 2017). Although extensive systems might appear to be less taxing on the environment in relation to resource use, waste treatment, and GHG emissions, scientific analysis has shown that intensive systems can actually reduce these outputs (Gerber et al., 2011; O'Brien et al., 2014).

Conversion of animal feed into edible animal products is always an efficiency concern in animal agriculture. And this is a particular concern in cases where animal feed may contain human-edible products, or animal feed is grown on land suitable for growing human food (Mottet et al., 2017). Approximately one-third of total cereal production is used to feed animals, and this is expected to rise further by 2030 (Makkar, 2017; Mottet et al., 2017). Currently 86 percent of global livestock feed is made of materials that are not consumed by humans, and ruminants play an important role in that they are uniquely able to convert human-inedible forages (e.g., leaves and grass) into high-quality protein and a variety of micronutrients (Mottet et al., 2017). Opportunities exist to increase this human-inedible proportion further, thereby decreasing the use of food-grade grains in both monogastric and ruminant diets.

Sustainability also encompasses animal welfare, and this is also a key component regarding consumer concerns. Unfortunately, there are few rigorous assessment criteria in use for scoring of animal welfare (Llonch et al., 2017). One definition states that animal welfare is "[when] animals are healthy and they have what they want" (Dawkins, 2017), while another definition from the World Organisation for Animal Health states that animal welfare is "[When] an animal is in a good state of welfare if (as indicated by scientific evidence) it is healthy, comfortable, well nourished, safe, able to express innate behavior, and if it is not suffering from unpleasant states such as pain, fear, and distress" (Terrestrial Animal Health Code,

OIE, Article 7.1.1). The Terrestrial Animal Health Code (the international standard for animal production) further outlines aspects of animal welfare for all of the major livestock species and includes objective assessments of behavior, morbidity, mortality, and reproductive performance, which can be used for taking into account animal well-being, thereby serving as the "scientific evidence" stated in the definition. However, there are few objective standards to assess many aspects of animal welfare, making quantitative evaluations problematic.

To leave sustainability goal evaluations to the political process or public opinion potentially exposes the process to subjective interpretation and political pressure from special-interest groups. There is a need to objectively evaluate the full sustainability implications of different agricultural systems. This would give the scientific community an opportunity to develop measures of product- and system-level performance to assess and compare the ability of different systems to sustainably meet the needs of both human and animal populations (Siegford et al., 2008). Likewise the holistic implications of utilizing mixed populations of animals that feed on wild grasses need to be objectively compared to the metrics associated with existing U.S. production systems to understand the food safety, environmental, animal health and well-being, worker safety and health, ranch viability, and food affordability trade-offs associated with these differing production systems.

An example of such rigorous assessment is one conducted by the Coalition for Sustainable Egg Supply (CSES), which formed to evaluate the sustainability of several laying-hen housing systems (CSES, 2018). The goal of the CSES is to provide scientifically based information on trade-offs related to varying housing systems by conducting holistic comparative research. CSES members represented various stakeholders, including food retailer companies, egg suppliers, animal welfare scientists, academic institutions, and government (USDA Agricultural Research Service) and nongovernmental organizations. This $6 million study examined "various laying hen housing systems and potential impacts on food safety, the environment, hen health and well-being, worker health and safety and food affordability, providing food system stakeholders with science-based information on sustainability factors to guide informed production and purchasing decisions" (Mench et al., 2016).

The study results demonstrated the complexity of addressing sustainability problems, as there were positive and negative aspects associated with each housing system that resulted in numerous trade-offs. Although the study was largely undertaken due to public pressure for cage-free hens because of animal welfare concerns, the cage-free system actually resulted in the highest rates of hen mortality and the worst indoor air quality, thus creating unexpected risks for both the animals and the workers.

Assessing the efficiency of livestock production systems likewise requires a holistic approach (Makkar, 2017). For example, Schader and colleagues (2015) explored the possibility of feeding animals only on human-inedible feedstuffs and found that it decreased the availability of livestock products globally by 53 percent, with a 91 percent decrease in meat from poultry and swine, and a 90 percent decrease in egg production compared with current levels of consumption. This underscores the important role of ruminants as consumers of human-inedible feedstuffs. However, the trade-off associated with removing high-energy concentrate feed from animal diets is that it increases the emission intensity of GHG production per unit of animal product.

Effective science communication of the results and trade-offs revealed in such studies will be essential to provide a factual basis for tackling sustainability issues in animal agriculture. Public opinion is not necessarily formed in response to objective scientific evidence, and especially when the issue involves ethical aspects, such as humane treatment of animals, the issue becomes much more complex, and simply reading or hearing the facts is not enough to sway majority outlook (Croney et al., 2012).

3. RESEARCH OPPORTUNITIES

Significant knowledge gaps and research opportunities exist in each of the sections below. To sustainably address the increasing animal protein demand will require advancing basic knowledge in core disciplines of animal science and will also entail considerable cross-disciplinary contributions.

3.1 Animal Genetics

There are new, hitherto unforeseen opportunities to accelerate genetic improvement of livestock by incorporating genomic information, advanced reproductive technologies, and precision breeding methods into conventional breeding and selection programs. Genomic selection alone has doubled the rate of genetic gain in the U.S. dairy industry since its introduction in 2009. The past decade has seen an explosion of genotyping and resequencing data that are currently being used to develop genomically enhanced breeding approaches in several industries (Weller et al., 2017). A large number of omics datasets (e.g., genomics, proteomics, metabolomics, and transcriptomics) have been and are being produced by researchers. The challenge now is to combine and analyze these big data to produce basic and applied knowledge on how best to genetically improve livestock populations for traits of interest. Hickey and colleagues (2017) proposed that genomic selection presents a unifying approach to bring plant and animal breeders together to deliver innovative "step changes" to the rate of genetic gain (see Figure 3-3).

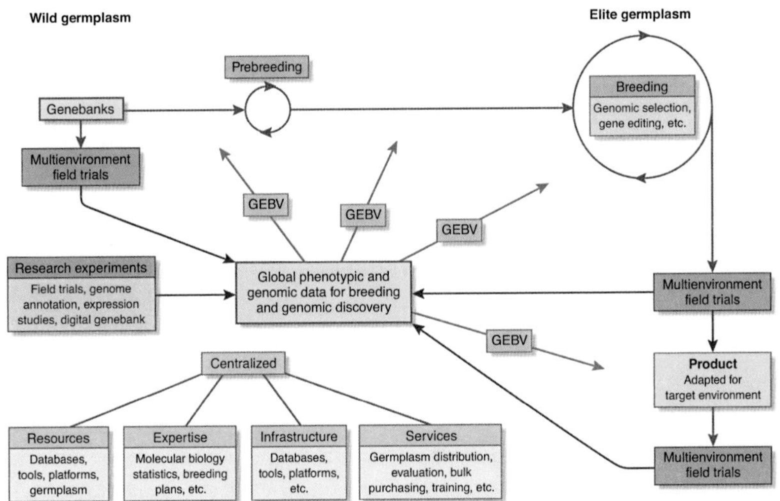

FIGURE 3-3 Capturing new opportunities to accelerate the pace of genetic gain based on efficient and targeted access to genetic diversity, coordinated phenotyping across environments, cost-effective sequencing, genomic prediction, and genome editing.
NOTE: GEBV = genomic estimated breeding value.
SOURCE: Hickey et al., 2017.

Sequence data from thousands of phenotyped animals may help uncover important quantitative trait polymorphisms that can be used to improve genomic predictions (VanRaden et al., 2017) or that can be introduced into breeds using genome-editing tools (Proudfoot and Burkard, 2017; Ruan et al., 2017). Using biological information on key single-nucleotide polymorphism loci obtained from sequencing projects, single base modifications can be precisely introduced into animal genomes to obtain a genotype with desired traits.

Novel breeding schemes involving multiple in vitro rounds of genomic selection, gene editing, gamete production, and fertilization could reduce by orders of magnitude both the generation interval and the genetic lag between nucleus and commercial populations. Such schemes can be envisioned based on exciting developments in embryonic stem cell technologies (Bogliotti et al., 2018) and surrogate sire/dam technology (Park et al., 2017; Taylor et al., 2017). This technology is poised to enable the development of a population of commercial animals that lack their own germline cells, but which carry transplanted gonial stem cells delivering the genetics from elite donor seedstock animals (Gottardo et al., 2018).

The main scientific challenge remains in how best to harmonize, combine, analyze, and utilize phenotypes, environmental, and omics information in conjunction with gene editing and advanced reproductive technologies. The goal would be a 10-fold increase in the rate of genetic improvement in livestock, poultry, and aquaculture populations by 2030. Additionally, there is a need to develop objective selection criteria to enable the incorporation of important components of sustainability, such as increased fertility, improved feed efficiency, functionality, and decreased susceptibility to disease into breeding objectives and animal breeding programs.

3.2 Animal Nutrition

Opportunities for research and progress abound for animal nutrition through the growing field of precision feeding and exploration of the microbiome, as well as through examination and implementation of novel feedstuffs. Precision feeding entails offering feed to each animal that is exactly tailored to the animal's needs. Currently, rations are formulated to be the least-cost ration that provides essential dietary requirements, which may result in overfeeding certain components such as protein. A recent paper examined the use of precision feeding stations to increase broiler flock uniformity by sequentially feeding chickens according to their individual body weight and needs (Zuidhof et al., 2017). Future innovations might address how feed rations can be more precisely formulated to maximize utilization efficiency and minimize negative environmental impact, and how technology could be used to more precisely deliver feed to animals on an individualized basis (Gerber et al., 2013). Also, the microbiome of animals is just beginning to be explored (O'Callaghan et al., 2016). Nutrient formulations could be combined with more information about the microbiome and its interactions with nutrients. As the human microbiome and associated research reveal connections between microbes and disease susceptibility, similar understandings will undoubtedly emerge concerning animals.

Finding novel feedstuffs for livestock could markedly enhance sustainability, as 14 percent of the global feedstock feed ration consists of human-edible feed materials (Mottet et al., 2017). Use of human-inedible resources—such as slaughterhouse and food wastes, by-products of biofuel production, leaf meals, seaweeds, and insect meals—could replace human-edible components of livestock diets (Rumpold and Schlüter, 2013; Makkar, 2017). Diverting food loss and waste to animal feed represents an obvious opportunity to replace grain-based feed with human-inedible resources. Plant breeding could also increase livestock production efficiency by (1) raising the feed crop yield per hectare (e.g., improved drought tolerance or nitrogen-use efficiency) and (2) improving the rate of feed conversion efficiency of vegetable calories into animal calories (e.g., altered digestibility or crop composition) such as reduced-lignin alfalfa (Van Eenennaam and Young, 2014). This will involve collaborations between plant breeders and animal nutritionists (see Chapter 2).

New electronic and digital tools may also enable better management of and feed conversion in ruminants grazing on grasslands and range. Virtual fences are a new foray into this area and could improve grassland productivity and protect sensitive areas from overgrazing (Umstatter, 2011).

3.3 Animal Health

Decreased production and death from disease are currently huge inefficiencies in the system, because nutrients and resources put into animal growth and well-being are minimized or completely lost. There are multiple knowledge gaps in animal health and thus many research areas in need of exploration.

Vaccine research to date has primarily focused on finding the right protein or engineered surrogate that will prompt a protective immune response when the animal encounters the specific pathogen in the field. Reverse vaccinology offers an exciting opportunity to use advances in genomics to predict good antigens for vaccine development and is currently being investigated by many researchers. For example, researchers have used reverse vaccinology to predict immunoprotective proteins for a tick vaccine (Andreotti et al., 2018) and leptospirosis vaccine candidates (Dellagostin et al., 2017). This approach was also used to identify vaccine candidates for a *Campylobacter* vaccine for broilers, an important food safety concern (Meunier et al., 2017). More targeted and individual-specific immune interventions based on multiple omics data may be the next breakthrough.

When a pathogen enters the body, there is a suite of upregulated host genes which informs the corresponding immune response, in a kind of signature response as part of the innate immune system. In this constellation of cascading cytokines, it is now known that some of these signaling molecules serve to protect the host and others may be triggered specifically by microbial genes to aid the organism's invasion (Davidson et al., 2015). Knowledge and dissection of this full complement of mediators in the innate immune response could serve to effectively inform better vaccine development that would be tailored for each invading agent. Such a schema could allow for earlier abrogation of the invading agent, rather than waiting for the acquired immune response to kick in, such as is induced by conventional vaccines.

If an infection cannot be prevented in an animal—in other words, there is no vaccine or it is not economically feasible to vaccinate—the next best tactic is to detect the illness at first blush. In most animal production facilities, animals are viewed every day by farmworkers and/or owners and are only seen by health professionals during periodic checks or when called to a farm for an assessment. When animals become visibly ill, they are already in the disease phase and are actively shedding the culpable microbes to their herd- or flock-mates. It would be ideal to be able to detect animals at first infection when they are in the incubation or prodromic stage (the stage of infection when symptoms begin to emerge but prior to full-blown disease becoming evident). Animals recognized at this stage could be removed from their cohorts and potential transmission drastically reduced.

An additional problem with animal health is diagnosing the disease so that appropriate intervention strategies can be introduced. In most cases, this involves first actually realizing that the animal is sick, then perhaps contacting an animal health professional for a visit, taking appropriate samples that will prove useful for diagnosing the specific disease occurring, and then relying on the local laboratory to run the appropriate tests. Current technologies, such as next-generation sequencing and MinION (a handheld DNA sequencer), are currently expensive and require high levels of expertise. With further research, these technologies could be converted into use at the penside rather than the laboratory. It is now possible to purify DNA from plants, animals, and microbes in under 30 seconds under field conditions (Zou et al., 2017). There is also ongoing work in the use of penside sensitive and specific biosensing systems for detecting animal diseases at the earliest moment, even before clinical signs are obvious (Vidic et al., 2017). The possibility of diagnosing a wide array of diseases while standing beside the animal could remove many time-consuming steps to diagnosis, allowing for much earlier and targeted therapies or control measures, thereby minimizing loses, animal suffering, and therapeutic antibiotic usage.

3.4 Animal Facility Design and Management for Sustainability and Animal Welfare

There are major research opportunities to study facility design and management for improving animal welfare and decreasing negative environmental consequences. Landmark studies by Dr. Temple Grandin (2012) provided insight into processing-facility design to reduce animal stress while at the same time improving worker safety. As a result of Grandin's work, more than half of the cattle slaughtered in the United States are processed through a curved, single-file chute, using nonslip flooring and adaptive lighting, resulting in more relaxed and calmer cattle walking to the slaughter site (*The Economist,* 2015). Opportunities exist to improve animal facility design by integrating sensors and electronic monitoring of animal health and well-being.

For layer hens, extensive research comparing different housing systems was conducted by CSES (previously described in this chapter). The study examined multiple parameters, including behavior, productivity, food safety, efficiency, and worker safety. This type of large holistic study provides the evidence base for system comparisons. Such research can be the impetus for research on how modification of facility design can improve welfare outcomes (Stratmann et al., 2015).

Rigorous, objective assessments of animal welfare need to be developed and applied. To date, few studies have examined animals' stress levels and/or senses of comfort and security. While the World Organisation for Animal Health has developed a robust set of parameters, most involve subjective behavioral or whole-animal observations. In some parts of the world, there has been progress in measuring internal biomarkers as a proxy for animal well-being. For example, cortisol can be quantitated noninvasively through saliva, milk, or hair (Casal et al., 2017; Tallo-Parra et al., 2017). The use of reliable biomarkers could help determine what types of housing facilities are best suited for the animals; facility design professionals could combine this information with their understanding of efficiency of operations, disinfection procedures, and worker safety and health.

Animal waste needs to be handled regardless of the animal housing system. Urine, feces, bedding material, and wash water are products of animal agriculture and are often destined for the waste stream. Designing programs that utilize or transform the waste into useful products are big opportunities for the research community. Some limited examples of successful research in this field include turning manure into solid paving materials, and it is becoming increasingly frequent to capture methane from manure for generating electricity (MacDonald et al., 2009; Fini et al., 2011). Continuing to find sustainable uses for animal wastes is a huge area of opportunity.

3.5 Precision Livestock Farming

Precision livestock farming (PLF) is a novel and growing technology with the aim of enhancing animal health and productivity by applying sensor and remote technologies. The use of PLF can enable individual-animal-targeted nutrition, health, and welfare (Rutter, 2012). Daniel Berckmans states that the objective of PLF is "to create a management system based on continuous automatic real-time monitoring and control of production/reproduction, animal health and welfare, and the environmental impact of livestock production" (Berckmans, 2014, p. 190). It is markedly multidisciplinary, requiring coordination among farmers, animal scientists, veterinarians, molecular biologists, immunologists, bioengineers, data scientists, and information technologists. Monitoring can be done remotely through sounds, sights, animal movements, and estimations of environmental parameters such as temperature, humidity, or air particulates.

Recent publications outline how PLF might be used in livestock systems to assess and control many aspects of animal lives (Rutter, 2012; Berckmans, 2014; Bocquier et al., 2014; Mottram, 2016). In the dairy cattle industry, the development of rumen wireless telemetry has enhanced metabolic disorder monitoring, and mastitis can be detected by combining conductivity and behavioral analysis with somatic cell scores that are predictive of disease (Mottram, 2016). Such technologies might offer an approach to reconcile the conflict that sometimes exists between animal welfare and efficiency.

Investigation of specific sounds using continuous monitoring has been examined in a preliminary way in two different livestock systems. Recording of sound emitted by broilers at various growth stages allowed for a correlation of sounds with optimal growth. This information could be used to investigate flocks when positive growth sounds are not heard, and also to modify the environment and feed to maximize the periods of these growth sounds being emitted (Fontana et al., 2015). In the cattle industry, recording of sounds in calves and labeling those noises specific for respiratory problems as a trigger for examination allowed for early identification of bovine respiratory disease (Vandermeulen et al., 2016).

Biosensors that detect physiological changes—such as increased lactate (indication of developing low milk yield in dairy cattle) or cortisol (indicating stress level)—have been developed that work on sweat, saliva, and tears (Weng et al., 2015). Applied to an animal, biosensors can serve as color-changing biotattoos to readily inform workers of impending problems. Information can be transferred via cell phone to create a timestamp as well as geographic coordinates that could be utilized for more effective local, regional, and national coordination of overall stress or developing disease. Prompt reaction to changes could facilitate isolation of infected animals to prevent spread, increase amount of weight gain, and decrease the number of sick animals that require therapeutic antibiotics. For PLF to reach its full potential, additional research into relevant bioresponses is needed. Currently, most of the point-of-care diagnostic development is focused on detection of specific pathogens (e.g., avian influenza). Although useful in monitoring flocks during outbreak periods, a more robust system would be able to detect early stress in an animal during the prodromic period. This will require intensive examination of the innate immune response and examination into which biomarkers are the most reliable for detection in blood, saliva, tears, or sweat. It will be necessary to detect specific algorithms and intervention points.

Phenotypes from sensors and PLF will also be important inputs to global efforts for collecting phenotypic and genomic data for breeding and genomic discovery programs. Combining large-scale sequence and varying phenotypic information from sensors into databases will be requisite to enable biology to inform breeding and feeding programs.

The biggest obstacle for PLF is the data science challenge of transforming multiple types of data from various sensors and sources into knowledge. This knowledge could then be used to accurately predict a genetically superior animal, an animal in distress or presenting disease symptoms, or an abnormal state that requires farmer intervention. There are two hurdles in this regard: (1) the data science problem associated with developing data-driven methods (e.g., deep learning) that can create predictions based on a variety of measurements coming from different sources, and (2) a translational issue of interpreting the data-driven predictions to produce prescriptive levels of understanding (see Chapter 6). This might range from developing a breeding value of an animal through to how and when a farmer should be notified that the intervention is required. Developing farmer-friendly, cost-effective plug-and-play PLF applications will require collaboration among animal scientists, agricultural engineers, and data scientists focused on a shared vision of animal-centered care.

3.6 Systems Analysis

In examining various production systems, or even animal-source foods versus plant-based or synthetic alternatives, it will be important to conduct a comprehensive life-cycle analysis (LCA) of advantages, disadvantages, and trade-offs. For instance, numerous studies have examined organic versus conventional production systems, but none have looked simultaneously at all relevant parameters involving an LCA to include environmental impact, animal welfare, and public health. To date, the most thorough studies in this arena have shown that there are evidently distinct advantages and disadvantages to each in the various categories of carbon footprint, water usage, and nutrient efficiency (van Wagenberg et al., 2017). For instance, although many consumers might choose grass-fed beef over cattle receiving conventional feed because of environmental concerns about the efficiencies of feeding grain to ruminants, the carbon footprint per unit of grass-fed beef can actually be considerably higher than cattle that are finished on concentrates (Capper, 2012).

There is a need to objectively evaluate the sustainability implications of different animal agricultural systems and protein source alternatives using a holistic evidence-based research approach. There are many questions that have been posed that deserve rigorous scientific analysis. Such questions include "What are LCA and nutritional implications of meat substitutes such as edible biomass grown from animal stem cells in culture, or plant-based meat imitation products?" Reducing animal agriculture may reduce GHG, but there are other direct and indirect impacts on other aspects of the food and agricultural system. A recent paper analyzed the environmental and nutritional implications of removing animals from U.S. agriculture and found a minimal overall reduction in GHG (2.6 percent), but significant potential impacts to human health through essential nutrient deficiencies (White and Hall, 2017).

Insects have been suggested as a future source of protein (Payne and Van Itterbeeck, 2017; Williams and Williams, 2017), and it will be worthwhile to explore the possibilities and implications of this protein source. A recent review by van Huis and Oonincx (2017) examined the LCA and environmental impacts of insect farming compared to livestock production. That study concluded that (1) less land and water is required, (2) GHG emissions are lower, (3) insects have high feed conversion efficiencies, (4) insects can transform low-value organic by-products into high-quality food or feed, and (5) certain insect species can be used as animal feed or aqua feed (van Huis and Oonincx, 2017). For instance, insects might be able to partially replace fish meal feed, which is becoming increasingly scarce and expensive and is important because of projected increases in aquaculture production (Williams and Williams, 2017).

Perhaps as important is communication of these findings to the broader public. The sustainability trade-offs associated with agricultural production systems are often not obvious. Marketers may focus on a single sustainability component of a production system without holistically addressing the trade-offs accompanying issues such as food safety and affordability, animal health and well-being, worker safety and health, and the environment.

In the United States, the Morrill Act of 1862 provided the basis for strong land-grant university research programs, and the subsequent Smith Lever Act of 1917 formed the extension infrastructure that would extend those research breakthroughs to farmers (Tokach et al., 2016). Extension programs could be better engaged in communicating agricultural sustainability research information to the general public as their target audience, using the findings of social scientists on the science of effective science communication, given the importance of the public being able to appreciate the nuances and trade-offs associated with different systems to sustainably meet the needs of both human and animal populations.

4. GAPS

There are scientific knowledge gaps that have prevented us from meeting the challenges so far and would need to be addressed in order to realize the opportunities. These are stated as defined basic or applied research questions below and address the areas of animal genetics, animal nutrition, animal health, animal facility design and management for sustainability and animal welfare, precision livestock farming, and systems analysis.

For animal genetics, research questions include (1) Can genetic and reproductive technologies be advanced and successfully combined to result in breeding schemes that achieve a 10-fold increase in the rate of genetic improvement in livestock, poultry, and aquaculture populations by 2030? (2) How can collaborative work with researchers in various fields (e.g., genetics, animal science, and animal health) help to develop reliable, objective selection criteria to enable genetic improvement in sustainability traits such as fertility, improved feed efficiency, welfare, and decreased susceptibility to disease in animal breeding programs?

For animal nutrition, research questions include (1) How can precision nutrition be utilized to ensure that each animal gets exactly what it needs? Would it be possible to combine nutrient information with growing data about animals' microbiomes to further the incorporation of exactness and efficiency? (2) Through collaborations between animal nutritionists and plant/food scientists, would it be possible to discover novel, nutritious, and suitable feedstuffs that are not human consumable in order to decrease animal agriculture's resource footprint?

For animal health, research questions include (1) What can be done to induce or amplify an animal's innate immune response to pathogens? What novel prevention and intervention therapeutics can be developed to interfere with the initial encounter and manipulate the innate response so that earlier protection is developed? Would it be possible to breed animals that are themselves resistant to specific pathogens? (2) What methods can be developed to more easily detect an infected animal prior to visual onset of clinical disease? What would be the role of biosensors and "wearable" technologies as inexpensive indicators? (3) To facilitate rapid response and containment, what penside tests can be developed to decrease the lag time between sample collection and laboratory diagnosis for transboundary animal diseases?

For animal facility design and management for sustainability and animal welfare, research questions include (1) How can animal housing and management be examined in a multidisciplinary way that would allow for high efficiency and productivity, combined with animal welfare and worker safety, while minimizing negative environmental impacts? (2) To assist in objective assessments of animal welfare, what parameters should be used to address welfare concerns in animal production systems? What are the appropriate objective modalities for measurement that can be incorporated (e.g., sound, sight, and movement) to help explore this area, particularly indicators of animals' subjective states? How can the social sciences help inform and translate scientific findings for consumers to make informed decisions?

For precision livestock farming, research questions include (1) How can sensors and data be used to transform the traditional systems of livestock farming into one that is more closely tailored to each animal's specific needs? (2) How should a "pilot farm" be designed to demonstrate the utility and efficiency of such a system that would provide continuous automatic real-time monitoring of production, animal health, welfare, and environmental impacts?

For systems analysis with regard to animal agriculture, research questions include (1) As the nation evaluates new systems to improve the efficiency of food production, how can life-cycle assessments be used to holistically examine and compare the trade-offs of proposed changes? Parameters for sustainability and acceptability should include food safety and nutrient composition, nutrition, environment, animal health and well-being, worker safety and health, and affordability.

5. EXAMPLES

Some possible applications of combining various breakthrough strategies to provide a systems approach are detailed below in futuristic examples.

5.1 Beef Cattle Biotattoo

A farmer raising beef cattle has a new group of calves. Cattle are housed together, and a monitoring system allows the farmer to visually detect, via a biotattoo in the ear, if the animal has ingested colostrum from the dam. A visual inspection shows the farmer that all the biotattoos are pink, indicating that this important antibody-laden milk has been successfully consumed, helping to protect the animal for the first few weeks of life. A biotattoo in the other ear indicates stress level, and a certain color indicates fever and/or production of acute-phase proteins, such as would be generated early in an infection. The farmer notes that two of the calves have a faint blue tinge to the biotattoo, so those calves and dams are brought in to an area for closer observation.

Before heading back to the house, the farmer activates the sound monitoring system, which records all sounds made by this group of calves and dams, and there are certain algorithms incorporated that will alert the farmer to the first coughs experienced in bovine respiratory disease and/or maternal distress. Any relevant sound algorithm that is activated alerts the farmer on a cell phone and, even in the middle of the night, the farmer is able to administer the appropriate remedy to the affected animals, saving lives and minimizing antibiotic use.

5.2 "Smart" Hog Facility

A state-of-the-art facility in the Midwest specializes in feeder hogs. Recently weaned pigs are brought in and given a thorough assessment by the consulting veterinarian. Sensors in all the animal rooms are tied to the air handling and lighting systems to ensure optimal temperature, humidity, and ambient light. A small number of workers monitor the animals remotely and are attuned to behavioral cues that the animals need to exert innate behaviors such as rooting or chewing; the workers can insert the requisite materials and/or activate access to an outdoor yard so that the animals can be fulfilled.

Weight gain and back fat are also measured remotely, and environmental parameters are modified to maximize lean muscle development. Fecal samples are periodically tested using inexpensive rapid assays for *Salmonella*. Because the microbiome of each pig is known, any carrier pig can quickly be identified, removed, and treated. Removal of sewage is conducted by robotics, and the manure is deposited in a methane transfer device that converts the energy and powers the entire facility. Transport to the slaughter plant is achieved through chutes and into trucks that mimic the pigs' environment for all of their senses.

5.3 Aquaponics in Elementary School

Elementary schools throughout a poor urban area, where fewer than 1 percent of children have ever been to a farm, are each equipped with an aquaponics unit, which sustainably combines fish production and vegetable growth, that has been incorporated into the science curriculum. Students are selected, on a rotating basis, to attend to the fish and the vegetables, and through their science classes learn about sustainable agriculture and the value of recycling. The fish and vegetables are periodically harvested and served in the school lunch program, with students helping in the preparation. After 2 years of this program, there is a noticeable increase in student interest in food science as a career, and several of the housing projects nearby have requested similar aquaponics units.

6. BARRIERS TO SUCCESS

Breakthroughs can only be accomplished if various governmental, societal, and funding concerns are also addressed, as currently there are obstacles to achieving what is envisioned through this report.

1. Public research funding levels have remained stagnant for many years, and are disproportionate to the economic contributions of

animal agriculture. The National Research Council report *Critical Role of Animal Science Research in Food Security and Sustainability* (NRC, 2015) recognized the underfunding of animal sciences and called for increased investments. Funding initiatives often omit research on food animals, focusing solely on the genetic improvement of agricultural crops. The National Association of State Departments of Agriculture has noted that "[t]his imbalance in support for animal science puts U.S. animal agriculture at a major disadvantage at a critical time when livestock, fish, and poultry producers are striving to improve sustainability and address global animal protein demands" (NASDA, 2014). Recent disease outbreaks such as porcine reproductive and respiratory syndrome and avian influenza in the United States underscore the need to develop science-based tools to prevent and mitigate the impacts of such outbreaks.

2. Big datasets are currently not stored or shared in a way that is useful for research or data scientists.
3. Many novel technologies are too expensive for commercial livestock producers and lack a user-friendly interface to facilitate their adoption.
4. There is little consumer understanding of the trade-offs associated with different production systems and agricultural innovations.
5. There is insufficient focus in current funding calls on bringing together the disparate disciplines needed to address complex problems using data science.

7. RECOMMENDATIONS FOR NEXT STEPS

Emerging technologies (such as genomics, genome editing, and biosensors) have transformative potential to advance knowledge in animal genetics, animal nutrition, and animal health. Funding mechanisms must entice and encourage data scientists, engineers, computer scientists, synthetic biologists, social scientists, and other nonagricultural disciplines to apply this biological knowledge to focus on developing innovative solutions to animal agriculture's pressing problems in a way that comports with societal expectations. A research strategy to enable the implementation of precision livestock farming using these technologies will require incentivizing the convergence of these disparate disciplines. Some high-priority research areas include the following:

1. Enable better disease detection and management using a data-driven approach through the development and use of sensing technologies and predictive algorithms.

2. Accelerate genetic improvement in sustainability traits (such as fertility, improved feed efficiency, welfare, and disease resistance) in livestock, poultry, and aquaculture populations through the use of big genotypic and sequence datasets linked to field phenotypes and combined with genomics, advanced reproductive technologies, and precision breeding techniques. The goal would be to enable a 10-fold increase in the rate of genetic improvement in livestock, poultry, and aquaculture populations by 2030.
3. Determine objective measures of sustainability and animal welfare, how those can be incorporated into precision livestock systems, and how the social sciences can inform and translate these scientific findings to promote consumers' understanding of trade-offs and enable them to make informed decisions.

REFERENCES

Alexandratos, N., and J. Bruinsma. 2012. *World Agriculture Towards 2030/2050: The 2012 Revision.* ESA Working Paper No. 12-03. Rome: Food and Agriculture Organization of the United Nations. Available at http://www.fao.org/docrep/016/ap106e/ap106e.pdf (accessed January 16, 2018).

Andreotti, R., P. F. Giachetto, and R. C. Cunha. 2018. Advances in tick vaccinology in Brazil: From gene expression to immunoprotection. *Frontiers in Bioscience (Scholar Edition)* 10:127-142.

Beal, T., E. Massiot, J. E. Arsenault, M. R. Smith, and R. J. Hijmans. 2017. Global trends in dietary micronutrient supplies and estimated prevalence of inadequate intakes. *PLoS ONE.* Available at https://doi.org/10.1371/journal.pone.0175554 (accessed May 30, 2018).

Bentley, J. 2017. *U.S. Trends in Food Availability and a Dietary Assessment of Loss Adjusted Food Availability, 1970-2014.* Economic Information Bulletin No. EIB-166. Washington, DC: U.S. Department of Agriculture, Economic Research Service.

Berckmans, D. 2014. Precision livestock farming technologies for welfare management in intensive livestock systems. *Revue Scientifique et Technique/Office International des Épizooties* 33(1):189-196.

Bocquier, F., N. Debus, and A. Lurette. 2014. Precision farming in extensive livestock systems. *Productions Animales* 27:101-111.

Bogliotti, Y. S., J. Wu, M. Vilarino, D. Okamura, D. A. Soto, C. Zhong, M. Sakurai, R. V. Sampaio, K. Suzuki, J. C. Izpisua Belmonte, and P. J. Ross. 2018. Efficient derivation of stable primed pluripotent embryonic stem cells from bovine blastocysts. *Proceedings of the National Academy of Sciences of the United States of America* 115(9):2090-2095.

Canfield, D. E., A. N. Glazer, and P. G. Falkowski. 2010. The evolution and future of Earth's nitrogen cycle. *Science* 330(6001):192-196.

Capper, J. L. 2011. The environmental impact of beef production in the United States: 1977 compared with 2007. *Journal of Animal Science* 89:4249-4261.

Capper, J. L. 2012. Is the grass always greener? Comparing the environmental impact of conventional, natural and grass-fed beef production systems. *Animals* 2(2):127-143.

Capper, J. L., R. A. Cady, and D. E. Bauman. 2009. The environmental impact of dairy production: 1944 compared with 2007. *Journal of Animal Science* 87(6):2160-2167.

Carpenter, S. R., N. F. Caraco, D. L. Correll, R. W. Howarth, A. N. Sharpley, and V. H. Smith. 1998. Nonpoint pollution of surface waters with phosphorus and nitrogen. *Ecological Applications* 8(3):559-568.

Casal, N., X. Manteca, D. Escribano, J. J. Ceron, and E. Fabrega. 2017. Effect of environmental enrichment and herbal compound supplementation on physiological stress indicators (chromogranin A, cortisol and tumour necrosis factor-α) in growing pigs. *Animal* 11(7):1228-1236.

Coffey, B., J. Mintert, S. Fox, T. Schroeder, and L. Valentin. 2005. *The Economic Impact of BSE on the US Beef Industry: Product Value Losses, Regulatory Costs, and Consumer Reactions.* Manhattan: Agricultural Experiment Station and Cooperative Extension Service, Kansas State University.

Croney, C. C., M. Apley, J. L. Capper, J. A. Mench, and S. Priest. 2012. The ethical food movement: What does it mean for the role of science and scientists in current debates about animal agriculture? *Journal of Animal Science* 90:1570-1582.

CSES (Coalition for Sustainable Egg Supply). 2018. *Welcome to the Coalition for Sustainable Egg Supply.* Available at http://www2.sustainableeggcoalition.org (accessed May 30, 2018).

Davidson, S., M. K. Maini, and A. Wack. 2015. Disease-promoting effects of type I interferons in viral, bacterial, and coinfections. *Journal of Interferon & Cytokine Research* 35(4):252-264.

Dawkins, M. S. 2017. Animal welfare and efficient farming: Is conflict inevitable? *Animal Production Science* 57(2):201-208.

Dellagostin, O. A., A. A. Grassmann, C. Rizzi, R. A. Schuch, S. Jorge, T. L. Oliveira, A. J. McBride, and D. D. Hartwig. 2017. Reverse vaccinology: An approach for identifying leptospiral vaccine candidates. *International Journal of Molecular Sciences* 18(1):E158.

The Economist. 2015. A jungle no more: How Temple Grandin's designs have reformed the meat industry. Available at https://www.economist.com/news/united-states/21671150-how-temple-grandins-designs-have-reformed-meat-industry-jungle-no-more (accessed January 16, 2018).

EPA (U.S. Environmental Protection Agency). 2018. Inventory of U.S. greenhouse gas emissions and sinks. EPA 430-R-18-003. Available at https://www.epa.gov/sites/production/files/2018-01/documents/2018_complete_report.pdf (accessed June 18, 2018).

Fini, E. H., E. W. Kalberer, A. Shahbazi, M. Basti, Z. You, H. Ozer, and Q. Zurangzeb. 2011. Chemical characterization of biobinder from swine manure: Sustainable modifier for asphalt binder. *Journal of Materials in Civil Engineering* 23(11):1506-1513.

Fontana, I., E. Tullo, A. Butterworth, and M. Guarino. 2015. An innovative approach to predict the growth in intensive poultry farming. *Computers and Electronics in Agriculture* 119:178-183.

Galasso, E., and A. Wagstaff. 2018. What cost childhood stunting? And what returns to programs combatting stunting? *Let's Talk Development.* World Bank. Available at http://blogs.worldbank.org/developmenttalk/what-cost-childhood-stunting-and-what-returns-programs-combatting-stunting (accessed January 20, 2018).

Gerber, P. J., T. Vellinga, C. Opio, and H. Steinfeld. 2011. Productivity gains and greenhouse gas emissions intensity in dairy systems. *Livestock Science* 139:100-108.

Gerber, P. J., A. N. Hristov, B. Henderson, H. Makkar, J. Oh, C. Lee, R. Meinen, F. Montes, T. Ott, J. Firkins, A. Rotz, C. Dell, A. T. Adesogan, W. Z. Yang, J. M. Tricarico, E. Kebreab, G. Waghorn, J. Dijkstra, and S. Oosting. 2013. Technical options for the mitigation of direct methane and nitrous oxide emissions from livestock: A review. *Animal* 7(Suppl. 2):220-234.

Gottardo, P., G. Gorjanc, M. Battagin, C. R. Gaynor, J. Jenko, R. Ros Freixedes, B. Whitelaw, A. Mileham, W. Herring, and J. Hickey. 2018. A strategy to exploit surrogate sire technology in animal breeding programs. Biorxiv 199893. Available at https://doi.org/10.1101/199893.

Grandin, T. 2012. Developing measures to audit welfare of cattle and pigs at slaughter. *Animal Welfare (South Mimms, England)* 21:351-356.

Griffin, D. 1997. Economic impact associated with respiratory disease in beef cattle. *Veterinary Clinics of North America: Food Animal Practice* 13(3):367-377.

Havenstein, G. B., P. R. Ferket, and M. A. Qureshi. 2003. Carcass composition and yield of 1957 versus 2001 broilers when fed representative 1957 and 2001 broiler diets. *Poultry Science* 82:1509-1518.

Hickey, J. M., T. Chiurugwi, I. Mackay, and W. Powell. 2017. Genomic prediction unifies animal and plant breeding programs to form platforms for biological discovery. *Nature Genetics* 49(9):1297-1303.

Ingram, J., P. Ericksen, and D. Liverman (Eds). 2010. *Food Security and Global Environmental Change*. London: Earthscan.

James, A. E., and G. H. Palmer. 2015. The role of animal source foods in improving nutritional health in urban informal settlements: Identification of knowledge gaps and implementation barriers. *Journal of Child Health and Nutrition* 4:94-102.

Llonch, P., M. J. Haskell, R. J. Dewhurst, and S. P. Turner. 2017. Current available strategies to mitigate greenhouse gas emissions in livestock systems: An animal welfare perspective. *Animal* 11(2):274-284.

MacDonald, J. M., M. Ribaudo, M. Livingston, J. Beckman, and W.-Y. Huang. 2009. *Manure Use for Fertilizer and for Energy: Report to Congress*. Administrative Publication No. AP-037. U.S. Department of Agriculture's Economic Research Service.

Makkar, H. P. S. 2017. Feed demand landscape and implications of food-not feed strategy for food security and climate change. *Animal* 4(12):1-11.

Mench, J. A., J. C. Swanson, and C. Arnot. 2016. The Coalition for Sustainable Egg Supply: A unique public-private partnership for conducting research on the sustainability of animal housing systems using a multistakeholder approach. *Journal of Animal Science* 94(3):1296-1308.

Meunier, M., M. Guyard-Nicodème, E. Vigouroux, T. Poezevara, V. Beven, S. Quesne, L. Bigault, M. Amelot, D. Dory, and M. Chemaly. 2017. Promising new vaccine candidates against *Campylobacter* in broilers. *PLoS ONE* 12(11):e0188472.

Mottet, A., and G. Tempio. 2017. Global poultry production: Current state and future outlook and challenges. *World's Poultry Science Journal* 73(2):245-256.

Mottet, A., C. de Haan, A. Falcucci, G. Tempio, C. Opio, and P. Gerber. 2017. Livestock: On our plates or eating at our table? A new analysis of the feed/food debate. *Global Food Security* 14:1-8.

Mottram, T. 2016. Animal board invited review: Precision livestock farming for dairy cows with a focus on oestrus detection. *Animal* 10:1575-1584.

NASDA (National Association of State Departments of Agriculture). 2014. Coalition letter to House and Senate Agriculture Appropriations Committees requesting FY 2015 funding for section 1433 of the Farm Bill. March 28. Available at http://www.nasda.org/letters-comments-testimony/coalition-letter-to-house-and-senate-agriculture-appropriations-committees-requesting-fy-2015 (accessed May 30, 2018).

NOAA (National Oceanic and Atmospheric Administration). 2015. *Fisheries of the United States, 2015: Per Capita Consumption*. Available at https://www.st.nmfs.noaa.gov/Assets/commercial/fus/fus15/documents/09_PerCapita2015.pdf (accessed January 16, 2018).

NRC (National Research Council). 2015. *Critical Role of Animal Science Research in Food Security and Sustainability*. Washington, DC: The National Academies Press.

O'Brien, D., J. L. Capper, P. C. Garnsworthy, C. Grainger, and L. Shalloo. 2014. A case study of the carbon footprint of milk from high-performing confinement and grass-based dairy farms. *Journal of Dairy Science* 97(3):1835-1851.

O'Callaghan, T. F., R. P. Ross, C. Stanton, and G. Clarke. 2016. The gut microbiome as a virtual endocrine organ with implications for farm and domestic animal endocrinology. *Domestic Animal Endocrinology* 56(Suppl.):S44-S55.

OIE (World Organisation for Animal Health). 2018. *Feeding the World by Controlling Animal Diseases.* Available at http://www.oie.int/for-the-media/editorials/detail/article/feeding-the-world-better-by-controlling-animal-diseases (accessed January 16, 2018).

Oladosu, G., A. Rose, and B. Lee. 2013. Economic impacts of potential foot and mouth disease agroterrorism in the USA: A general equilibrium analysis. *Journal of Bioterrorism & Biodefense* S12:001. doi: 10.4172/2157-2526.S12-001.

Park, K. E., A. V. Kaucher, A. Powell, M. S. Waqas, S. E. Sandmaier, M. J. Oatley, C. H. Park, A. Tibary, D. M. Donovan, L. A. Blomberg, S. G. Lillico, C. Bruce, A. Whitelaw, A. Mileham, B. Telugu, and J. M. Oatley. 2017. Generation of germline ablated male pigs by CRISPR/Cas9 editing of the *NANOS2* gene. *Scientific Reports* 7:40176.

Payne, C. L. R., and J. Van Itterbeeck. 2017. Ecosystem services from edible insects in agricultural systems: A review. *Insects* 8(1):24.

Proudfoot, C., and C. Burkard. 2017. Genome editing for disease resistance in livestock. *Emerging Topics in Life Sciences* 1:209-219.

Ruan, J., J. Xu, R. Y. Chen-Tsai, and K. Li. 2017. Genome editing in livestock: Are we ready for a revolution in animal breeding industry? *Transgenic Research* 26:715-726.

Rumpold, B. A., and O. K. Schlüter. 2013. Potential and challenges of insects as an innovative source for food and feed production. *Innovative Food Science & Emerging Technologies* 17:1-11.

Rutter, S. 2012. A "smart" future for ruminant livestock production. *Cattle Practitioner* 20:186-193.

Schader, C., A. Muller, N.-H. Scialabba, J. Hecht, A. Isensee, K. H. Erb, P. Smith, H. P. Makkar, P. Klocke, F. Leiber, P. Schwegler, M. Stolze, and U. Niggli. 2015. Impacts of feeding less food-competing feedstuffs to livestock on global food system sustainability. *Journal of the Royal Society, Interface* 12(113):20150891.

Siegford, J. M., W. Powers, and H. G. Grimes-Casey. 2008. Environmental aspects of ethical animal production. *Poultry Science* 87(2):380-386.

Stratmann, A., E. K. F. Froehlich, and S. Gebhardt-Henrich. 2015. Modification of aviary design reduces incidence of falls, collisions and keel bone damage in laying hens. *Applied Animal Behaviour Science* 165:112-123.

Tallo-Parra, O., E. Albanell, A. Carbajal, L. Monclus, X. Manteca, and M. Lopez-Bejar. 2017. Prediction of cortisol and progesterone concentrations in cow hair using near-infrared reflectance spectroscopy (NIRS). *Applied Spectroscopy* 71(8):1954-1961.

Taylor, L., D. F. Carlson, S. Nandi, A. Sherman, S. C. Fahrenkrug, and M. J. McGrew. 2017. Efficient TALEN-mediated gene targeting of chicken primordial germ cells. *Development* 144(5):928-934.

Tokach, M. D., B. D. Goodband, and T. G. O'Quinn. 2016. Performance-enhancing technologies in swine production. *Animal Frontiers* 6(4):15-21.

Umstatter, C. 2011. The evolution of virtual fences: A review. *Computers and Electronics in Agriculture* 75(1):10-22.

UN DESA (United Nations Department of Economic and Social Affairs). 2015. *World Population Projected to Reach 9.7 Billion by 2050.* New York: UN DESA. Available at http://www.un.org/en/development/desa/news/population/2015-report.html.

USDA-APHIS (U.S. Department of Agriculture's Animal and Plant Health Inspection Service). 2016. *2016 HPAI Preparedness and Response Plan.* Available at https://www.aphis.usda.gov/animal_health/downloads/animal_diseases/ai/hpai-preparedness-and-response-plan-2015.pdf (accessed January 16, 2018).

USDA-ERS (U.S. Department of Agriculture's Economic Research Service). 2018a. *Animal Products.* Available at https://www.ers.usda.gov/topics/animal-products (accessed January 16, 2018).

USDA-ERS. 2018b. *U.S. Agricultural Trade Data Update.* Available at https://www.ers.usda.gov/data-products/foreign-agricultural-trade-of-the-united-states-fatus/us-agricultural-trade-data-update/#Latest%20U.S.%20Agricultural%20Trade (accessed May 1, 2018).

USDA-FAS (U.S. Department of Agriculture's Foreign Agricultural Service). 2018. Production, supply, and distribution online database. Available at https://www.fas.usda.gov/databases/production-supply-and-distribution-online-psd (accessed June 17, 2018).

Valin, H., R. D. Sands, D. van der Mensbrugghe, G. C. Nelson, H. Ahammad, E. Blanc, B. Bodirsky, S. Fujimori, T. Hasegawa, P. Havlik, E. Heyhoe, P. Kyle, D. Mason-D'Croz, S. Paltsev, S. Rolinski, A. Tabeau, H. van Meijl, M. von Lampe, and D. Willenbock. 2014. The future of food demand: Understanding differences in global economic models. *Agricultural Economics* 45:51-67.

Van Eenennaam, A. L., and A. E. Young. 2014. Prevalence and impacts of genetically engineered feedstuffs on livestock populations. *Journal of Animal Science* 92:4255-4278.

van Huis, A., and D. G. A. B. Oonincx. 2017. The environmental sustainability of insects as food and feed. A review. *Agronomy for Sustainable Development* 37:43.

van Wagenberg, C. P. A., Y. de Haas, H. Hogeveen, M. M. van Krimpen, M. P. M. Meuwissen, C. D. van Middelaar, and T. B. Rodenburg. 2017. Animal Board invited review: Comparing conventional and organic livestock production systems on different aspects of sustainability. *Animals* 11(10):1839-1851.

Vandermeulen, J., C. Bahr, D. Johnston, B. Earley, E. Tullo, I. Fontana, M. Guarino, V. Exadaktylos, and D. Berckmans. 2016. Early recognition of bovine respiratory disease in calves using automated continuous monitoring of cough sounds. *Computers and Electronics in Agriculture* 129:15-26.

VanRaden, P. M., M. E. Tooker, J. R. O'Connell, J. B. Cole, and D. M. Bickhart. 2017. Selecting sequence variants to improve genomic predictions for dairy cattle. *Genetics Selection Evolution* 49(1):32.

Vidic, J., M. Manzano, C. M. Chang, and N. Jaffrezic-Renault. 2017. Advanced biosensors for detection of pathogens related to livestock and poultry. *Veterinary Research* 48(1):11. doi: 10.1186/s13567-017-0418-5.

Weller, J. I., E. Ezra, and M. Ron. 2017. Invited review: A perspective on the future of genomic selection in dairy cattle. *Journal of Dairy Science* 100:8633-8644.

Weng, X., L. Chen, S. Neethirajan, and T. Duffield. 2015. Development of quantum dots-based biosensor toward on-farm detection of subclinical ketosis. *Biosensors and Bioelectronics* 72:140-147.

White, R. R., and M. B. Hall. 2017. Nutritional and greenhouse gas impacts of removing animals from US agriculture. *Proceedings of the National Academy of Sciences of the United States of America* 114(48):E10301-E10308.

Williams, D. D., and S. S. Williams. 2017. Aquatic insects and their potential to contribute to the diet of the globally expanding human population. *Insects* 8:72.

World Bank. 2013. Fish to 2030: Prospects for fisheries and aquaculture. World Bank Report Number 83177-GLB. Available at http://www.fao.org/docrep/019/i3640e/i3640e.pdf (accessed June 27, 2018).

Zou, Y., M. G. Mason, Y. Wang, E. Wee, C. Turni, P. J. Blackall, M. Trau, and J. R. Botella. 2017. Nucleic acid purification from plants, animals and microbes in under 30 seconds. *PLoS Biology* 15(11):e2003916.

Zuidhof, M. J., M. V. Fedorak, C. A. Ouellette, and I. I. Wenger. 2017. Precision feeding: Innovative management of broiler breeder feed intake and flock uniformity. *Poultry Science* 96(7):2254-2263.

4

Food Science and Technology

1. INTRODUCTION

As the U.S. food system has evolved, advances in science and technology have helped to provide a huge variety of foods that are safe, convenient, inexpensive, distributed widely, and available year round. Individuals representing many disciplines—microbiology, chemistry, engineering, processing, packaging, sensory science, and nutrition, among others—work under the umbrella of food science to support the integrity of the food supply (Floros et al., 2010). In addition, food scientists collaborate with other disciplines (e.g., agronomists, biotechnologists, material scientists, economists, and social/behavioral scientists) to address problems in the broader "food system," with the ultimate purpose of transforming raw, frequently inedible, and, in some cases, unsafe agricultural commodities into safe, nutritious, high-quality foods that are accepted and valued by consumers (see examples in Box 4-1). Much of this is accomplished by food processing, defined as any intentional change to a food occurring between the point of origin and availability for consumption (Floros et al., 2010). Food is processed for many different purposes and, overall, processing results in improved product characteristics such as safety, shelf life, quality, sensory attributes, and nutritional value. In more recent years, consumers have demanded additional product features such as convenience and variety to their food choices, and they expect greater transparency about the origins of their food and the type of processes utilized in manufacturing a product. New trends such as online food shopping and the use of food-on-demand services

BOX 4-1
Examples of the Contribution of Food Science and Technology to Improving the Food Systems

Frozen Foods

For centuries, inhabitants of cold climates have used the natural winter frosts as a preservation approach to protect foods from microbial or enzymatic deterioration. Following early experience with ice making in the 19th century, technology developed rapidly and by the end of the 19th century, cold air freezing plants were being used to produce frozen poultry in Europe, which was being traded between countries. In the 1920s, Clarence Birdseye developed flash freezing of food, which produced a product of much higher quality; this technology was investigated by the military during World War II. As an evolving society embraced convenience and variety, and the food industry recognized financial benefits such as enhanced capacity and extended shelf life, freezing technologies were increasingly embraced in the decades that followed. This has led to innovations such as frozen meals available in groceries to the recent emergence of the food-on-demand businesses that provide door-to-door delivery of ingredients and products by overnight shipment.

Food Fortification

Food fortification or enrichment can be defined as the intentional process of adding micronutrients to food. A historical example is the addition of iodine to salt, which began in the 1920s because the lack of natural iodine was resulting in a high prevalence of goiter in the population. Among the many examples of improving health via fortification is the U.S. mandatory fortification of all enriched grain products with folic acid, which has prevented thousands of neural tube defect incidences per year. Developments in nanotechnology, specifically nanoencap-

allow for even greater individualization in consumer choice, preference, and demand.

Attaining a food supply that provides safe, healthy, appealing, and affordable foods is the shared responsibility of food and allied industries, local, state, and federal governments, and researchers and educators in academic institutions, along with consumers through their food choices and practices. Most of the necessary research and development (R&D) work to launch new commercial products is naturally initiated and conducted by the private sector. However, investigating overarching concepts in the food sciences, and solving universal, crosscutting problems, is frequently tackled with basic and applied scientific research that is conducted at public and private universities and in government laboratories. Although different stakeholder groups contribute to the funding and intellectual enterprise of sulation, are providing means by which to improve our fortification capabilities by, for instance, protecting sensitive bioactive compounds from acidic environments, exogenous enzymes, or thermal degradation; enhancing solubility; and/or improving release or bioavailability (Singh et al., 2017). New investigations are focusing on biofortification, the use of agronomic practices, conventional plant breeding, and modern biotechnology to increase nutrient levels in crops during plant growth (e.g., iron biofortification of rice, beans, or sweet potatoes; zinc biofortification of wheat; provitamin A carotenoid biofortification of sweet potatoes or maize) (WHO, 2016).

Upcycling: Reuse of By-Products from Food Processing

A classic example of early upcycling was the utilization of sweet whey, a by-product of cheese making, as an ingredient in various food products. Considered a waste by early cheese producers, whey was pumped into rivers and streams, which created dead zones in the ecosystem because of overgrowth of algae (Marwaha and Kennedy, 1988). When this practice was banned, whey began to be used in various capacities, first as animal feed, then as a filler, and eventually as an ingredient of "health food" products. In addition, cream skimmed from whey was used to make whey butter, an ingredient in butter-flavored food.

Upcycling has gained significance among food manufacturers not only as a lucrative process but more importantly as a part of their resource efficiency and sustainability management plans. Other historical successes of upcycling include the use of fishing bycatch to produce value-added products such as surimi. A modern example of the need for upcycling is the mounting quantities of acid whey being produced as a result of the recent popularity of Greek-style yogurt. Researchers are investigating ways to upcycle this new food processing by-product.

the agricultural and food-related activities, historically the research efforts have been largely supported by both public and private funds. Between 1970 and 2008, the public contribution was relatively stable at about 50 percent of the total private- and public-sector R&D funding (Clancy et al., 2016). Recently, however, the source of funding has shifted. During the period 2008 to 2013, real private investment in R&D in the agricultural and food sector rose sharply (up by 64 percent), while real public investment fell by 20 percent. Private funding has dominated R&D in food manufacturing (Clancy et al., 2016). Public support for human nutrition research has increased over the past several decades. The nutrition research includes nutrition through the life cycle, health (disease, metabolism, and metabolic mechanisms), and food science (monitoring, education, and policy; and supplements). However, the portfolio of research has changed with increased funding from the U.S. Department of Health and Human Services and decreased support from the U.S. Department of Agriculture. The shift has affected the type of problems addressed through federal support as well as mechanism (shifting from formula funds to nonformula extramural support). From 1985 to 2009, the federal share of research funding for food sciences (food processing, preservation, and other food-related technologies) decreased from 10 to 4 percent of the total funding for nutrition research (Toole and Kuchler, 2015).

This chapter identifies important challenges faced by the postharvest food sector in making progress toward meeting future demands for a safe, nutritious, sustainable, and affordable food supply for all. It also identifies emerging opportunities, largely as a consequence of scientific and technological developments, to address these challenges, along with gaps and barriers. Concrete illustrative examples of these emerging opportunities are provided. The chapter does not address the cost and social implications of these technological advances, including factors that may limit access to new products or processes (e.g., production scale, location, or consumer resources), although it is recognized that these factors are important drivers of their ultimate adoption. Chapter 9 considers some of the socioeconomic considerations related to the scientific innovations.

2. CHALLENGES

Factors such as population growth, more variable weather cycles, and globalization, among others, have changed and continue to dramatically change our food system. Supply networks now offer greater consumer choice over a wide variety of products through large, interconnected markets. However, many challenges to the system have emerged. The committee identified two general challenge areas that need to be addressed over the next 20 years using the newest scientific and technological breakthroughs.

2.1 Challenge 1: Develop High-Quality, Nutritious Foods Produced and Distributed in a Sustainable Manner to Meet the Needs and Demands of a Diverse Consumer Population

The essential role of food is maintaining human life and health. Food promotes health because it contains nutrients that are necessary to provide energy, meet physiological needs and functions, and prevent chronic diseases. As mentioned in Chapter 1, this report does not address research efforts devoted to understanding the association between human nutrition and health, although it should be noted that this continues to be an important area of future research. Indeed, the increased recognition of the complex, and often personalized, interactions between agricultural production, food, nutrients, and human health begs for research to improve our understanding of food and nutrient metabolism and their relationship to diet and health. Findings from this type of research could lead to more healthful foods and better diets in general, and those in accordance with the needs of specific consumer subpopulations.

It is important to recognize that humans eat foods, not nutrients, and so foods must be both nutritious *and* appealing. Sensory attributes are among the most important drivers of food consumption preferences (Lusk and Briggeman, 2009). The holistic sensory experience is complex, and there is an implicit causal chain of events from sensation, to experiencing pleasure, to food intake. Sensory is not only impacted by the complexity of food components from macromolecules to ingredients to formulation; there is emerging science indicating that human genetic variability plays a major role in the way individuals experience foods. Understanding the interactions between the food chemical composition and the consuming human is critical to developing products that meet consumer preferences for flavor and appearance while delivering nutrition and health benefits.

In addition to consumer appeal and healthfulness, consumers' eating preferences are driven by many social, behavioral, and psychological factors (Lusk and McCluskey, 2018). For some consumers, ethical and environmental concerns may dominate their preferences (e.g., vegetarian protein substitutes for animal products; insects used as a source of protein); for others, place of origin and local sourcing are predominant considerations; and in other cases, perceived risks weigh heavily in food choice (e.g., choice of organic options, avoidance of genetically engineered foods or other new technologies). Improved understanding of the influence of social, behavioral, and psychological factors on the development and role of these influences is necessary, particularly as consumers are faced with choices about products developed with new technologies for some of which there is conflicting evidence on risks and benefits. One relevant ethical issue is that of consumer behavior around food loss and waste, given that 30-40 percent of the food produced in the United States is wasted, largely at the retail and consumer stages (Gunders, 2012; Buzby et al., 2014; Bellemare et al., 2017). Food supply chain participators have joined forces in initiatives to reduce waste (e.g., changing product labeling policies) but important technological innovations can be added to these efforts, including development of ways to increase product quality, shelf life, and/or safety. Other challenges are best addressed through focus on a systems approach and behavioral changes (see Conrad et al., 2018, for an example of the challenge of improved diet quality being associated with increased food waste and greater amounts of water and pesticide use).

2.2 Challenge 2: Protect the Integrity and Safety of the Global Food Supply Chain

An increasingly globalized and highly networked food supply chain has made it more challenging to protect food from intentional and unintentional microbial and chemical contamination. Although regulatory and surveillance systems are arguably better than they were 25 years ago, in many ways our current food safety system still lacks sophistication and is not nimble enough to respond swiftly when a critical issue arises.

Assurance of food safety relies on preventing contamination or removing/inactivating the contaminant if it occurs along the chain. Large amounts of food safety data are currently being collected from farm to fork, but those data can be somewhat crude (e.g., visual inspection of poultry carcasses along a processing conveyor rather than instrumentation measurements) and when measurements are made, they are simple (i.e., nonquantitative) and delayed; certainly, they most often provide only a snapshot in time. New technologies are making it possible to obtain more sophisticated data, sometimes collected continuously and/or in real time. If more precise, accurate, faster, and less expensive technologies were applied to food protection, testing could occur more often to facilitate the detection of infrequent contamination events, and to more rapidly manage and respond to food safety incidents. For example, the availability of very rapid and sensitive ways to detect harmful biological agents or chemical contaminants would result in a safer food supply, especially if detection occurred before the contaminants were widely dispersed as ingredients or through products entering the retail food system. This would be particularly the case if the methods were easy to apply and inexpensive. Identifying the most relevant data and points of collection and intervention are key to effective and integrated data systems. Field deployability would allow detection technologies to touch every phase of the farm-to-fork continuum.

When a contaminated product enters the market, or an outbreak occurs, we currently rely on piecemeal systems to perform epidemiological investigations, trace back, and trace forward, meaning public health risk remains elevated for extended periods of time, until the right information has been obtained and synthesized. A thorough and integrated data communication and management system that includes all steps in the supply chain would greatly aid traceability and reduce the public health impact of food safety events, particularly in the case of larger processors, distributors, and retailers. As stated above, technological advances over the past few decades have opened the door to faster, more accurate, and more relevant data collection in food safety. When married to algorithms that assess risk and costs and benefits, it is possible to prevent contaminated products from entering the food chain or, if they do, prevent their further distribution and consumption in a matter of minutes or hours, not days or weeks.

There is also a need to ensure that best practices to maintain food quality are being adhered to throughout the food supply and distribution channels. For instance, data from biochemical analysis can be used to ensure that product traits such as appearance, flavor, or nutritional value are maintained. An integrated system that mapped the flow of products and ingredients, and transferred information about food quality throughout food distribution, would improve efficiency and integrity by contractors all through the supply chain and increase consumer trust. Better assurance of food quality will also aid in optimizing resource efficiencies in the system and ultimately reduce food loss and waste through improved ingredient flow and increased product shelf life.

3. SCIENTIFIC OPPORTUNITIES

3.1 Opportunity 1: Omics Technologies

The recently coined term foodomics refers to the use of "omics" technologies and data as they relate to the discipline of food science (Capozzi and Bordoni, 2013; Andjelkovic et al., 2017). For example, integrated analytical approaches in food chemistry and analysis can be used to increase our understanding of food composition at the molecular and even atomic levels. It is now possible to produce food "fingerprints" of chemical composition, information that is relevant to safety, quality, authenticity, security, and nutritional value (Gallo and Ferranti, 2016). Beyond food fingerprinting, omics technologies provide a means to detect, quantify, and characterize individual metabolites or combinations thereof. This is opening doors to development of improved bioactive absorption and delivery systems, and better colors and flavors, to name just a few of the applications (Gallo and Ferranti, 2016). These technologies are also particularly useful in identifying relevant volatile compounds that may serve as markers of product freshness (Wojnowski et al., 2017), for improving food quality, and for ultimately reducing food loss and waste. They may also identify molecular targets (analytes) during the development of advanced detection methods for harmful microbes, chemicals, and toxins, and therefore further improve food safety. Identification of novel biorecognition molecules used to capture and detect key analytes will make it easier to perform analyses on very complex sample matrices, a long-time obstacle to the application of advanced analytical methods to foods. Production of increasingly miniaturized analytical equipment (i.e., infrared, ultraviolet, mass spectrometry, and nuclear magnetic resonance [NMR] spectroscopy), some of which can automate sampling and analysis for real-time biochemical measurement, offers opportunities for exquisitely sophisticated chemical analysis that may become field deployable.

The combined use of omics technologies, bioinformatics, and advanced analytical methods provides innovative means by which scientists can explore interactions between systems. In nutrition, for instance, applying omics techniques to human genetics, physiological status, the gut microbiome, and food composition can lead us closer to integrated personalized nutrition (Grimaldi et al., 2017; Kaput et al., 2017) (see Box 4-2). In sensory science, where we know that the flavor experience is multimodal, omics techniques can be used to characterize genetic and metabolic differences in consumer perception of flavor, allowing for a better understanding

BOX 4-2
Examples of Precision Nutrition

Personalized (or precision) nutrition is a customizable approach to dietary management, providing tailored nutrition advice that is relevant to the varied needs of a specific individual or group of individuals. While this has more or less been the approach of nutritionists for decades, the availability of advanced omics techniques has allowed for better understanding of the relationship between food, genetics, physiology, lifestyle, and other factors. When these data are integrated and appropriately analyzed for an individual person or a group, it is possible to predict physiological responses, manage or even prevent disease states, and balance physical with emotional needs, among others.

An example of using omics and nanotechnology for precision nutrition is a measurement of vitamin B_{12} levels in blood. Vitamin B_{12} is an essential nutrient, and low serum levels have been associated with adverse metabolic health profiles, such as insulin resistance and cardiovascular disorders. Genome-wide association studies have identified more than 12 genetic variants associated with serum vitamin B_{12} levels. One variant of the human gene *FUT2*, encoding a protein involved in the attachment of the bacterium *Helicobacter pylori* to the gastric mucosa, is associated with low vitamin B_{12} levels in the blood. A variant of the human gene *TCN2* is associated with both low serum vitamin B_{12} and high homocysteine levels. Mutations in several other genes, for example, *TCN3*, *GIF*, *CUBN*, and *AMN*, have been identified as causes of various vitamin B_{12} absorption deficiency syndromes. A mobile platform for the rapid analysis of blood vitamin B_{12} levels has been recently developed. The NutriPhone technology, which is composed of a smartphone accessory along with an app, is a prototype that quantifies vitamin B_{12} at very low physiological levels using an innovative test strip that was built using nanotechnology based on silver amplification technology. Its utility has recently been demonstrated in a small study on human subjects (Lee et al., 2016) and it is a good illustration of the ability to harness advanced technology toward the ultimate goal of personalized nutrition.

of what drives food choice. When this information is used along with food fingerprinting, it becomes possible to design and produce food having ideal health benefits with greater consumer appeal.

3.1.1 Gaps

Individual omics technologies focus on one aspect or component of a much larger system. In a health care setting, genomics can be used for genetic fingerprinting, metabolomics for metabolic profiling, sequencing and bioinformatics for elucidating characteristics of the microbiome. For a particular food, various omics techniques can be used to determine its nutrient composition, sensory characteristics, and microbiological profile. Each of these individual analyses provides characterization of what is going on in a patient or a product and constitutes a subsystem. However, to understand the entire person or product, there is also the need to elucidate how these subsystems interact with one another, forming a system of systems. For instance, most chronic diseases (e.g., diabetes and cardiovascular disease) are complex, with diet being only one contributing factor. For such diseases, there are significant gaps in knowledge about interactions between genes, diet, other behaviors (e.g., exercise and stress), and social and cultural factors, among others. Having the full scientific capabilities to understand the interactions and identify the key determinants of any particular illness or trait has yet to be realized (see Box 4-3).

3.2 Opportunity 2: Sensor Technologies

According to a recent study, the most common reasons given by consumers for discarding food were concerns about its safety and the willingness to consume only the freshest product (Neff et al., 2015). Having a technology that can "sense" product safety, quality, and/or freshness, preferably in real time, will deliver critical information to processors, distributors, and consumers, potentially resulting in better decisions about safety and food waste. Such technologies ideally would have features such as high sensitivity and specificity of analyte detection, low cost, small footprint, reliability, short time to result, and be field deployable and adaptable, among others.

Sensors are devices that detect or measure physical, chemical, or biological properties and then record, indicate, or respond to those results. Biosensors in particular are analytical devices that combine a biological component with a physicochemical detector. The biologically derived component is a material or biomimetic compound that interacts, binds, or otherwise recognizes the analyte to be detected. Increasingly, these are being identified using various omics methods (see the section above). The

BOX 4-3
Gaps in Understanding a System of Systems: Determinants of Chronic Diseases

The gap in understanding systems within systems is clearly illustrated in the area of diet, food, and nutrition where one can think of various systems—genetic, epigenetic, and nutrient–gene interactions—driving the variation in risk of diseases and resulting in individual nutritional requirements. Figure 4-1 shows the relationships among various systems as they drive the phenotypic variation in cardiovascular disease (CVD), an example of a complex chronic disease (NASEM, 2018). Although dozens of single-nucleotide polymorphisms (SNPs) and combinations of genes (genetic risk scores [GRSs]) have been associated with higher risk of CVD phenotypes in several genome-wide association studies, the role of diet in these associations is poorly understood. The figure shows how a variety of nutrient components (e.g., a food, a nutrient, or a food group) can modify the genetic risk of a disease by interacting with a genetic component (SNP, GRS, or whole-genome next-generation sequencing). Although we can expect enhancements to the predictive value of SNPs and GRSs for CVD events and the identification of patients at increased genetic risk, there are gaps in the evidence for designing personalized diets for optimizing the prevention of CVD.

Understanding such a complex system and the number of parameters to be estimated within it—complex interactions and nonlinear dose-response relationships among all the nutrients in the food supply, all the microbes in the human gut, all the polymorphisms in the human genome (Grimaldi et al., 2017; Kaput et al., 2017), and additional environmental factors—necessitates not only the generation of data but also the data science technologies to store, curate, analyze, and share the large volumes of data.

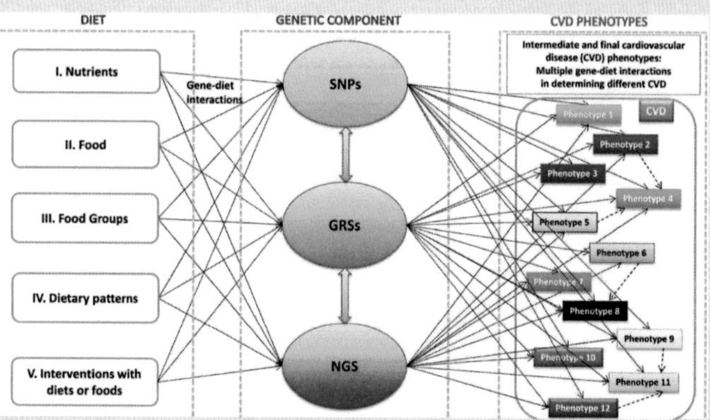

FIGURE 4-1 Complexities of nutrigenomics, using cardiovascular disease (CVD) phenotypes as the endpoints.
NOTE: GRS = genetic risk score; NGS = next-generation sequencing; SNP = single-nucleotide polymorphism.
SOURCE: Corella et al., 2017.

interaction between the biological element and the analyte results in a signal; a detector element physicochemically transforms (transduces) that signal, and frequently amplifies it into a form that is readily measurable and sometimes quantifiable. There are many types of transducers, such as electrochemical, optical/visual, and mass based (Vigneshvar et al., 2016; Alahi and Mukhopadhyay, 2017). Table 4-1 provides a summary of some common biosensor technologies.

Nanomaterials are increasingly used as components of biosensors and can serve a variety of functions, including as immobilization supports, for signal amplification, as alternatives to enzyme labels ("nanozymes"), and to aid in signal generation and quenching (Rhouati et al., 2017). In most cases, the choice to use nanomaterials is founded on the desire

TABLE 4-1 Summary of Various Biosensors with Their Advantages and Limitations

Method of Detection	Advantages	Limitations	Cost	References
Optical methods	Sensitivity is high, can detect almost in real time, and detection system is label free	Cost is very high	High	Mandal et al., 2011; Zhang, 2013
Electrochemical methods	Requires large quantity of sample numbers, might be automatic, and detection system is label free	Specificity is low, not suitable for low sensitivity, and needs a lot of washing steps	Low	Mandal et al., 2011; Zhang, 2013
Mass-based methods	Cheaper than other methods, easy operation, able to detect in real time, and detection is label free	Specificity and sensitivity are low, requires long incubation time, and regenerating crystal surface is problematic	Low	Mandal et al., 2011; Zhang, 2013
Nanomaterial-based sensors	User-friendly measurement, and measurement can be done in real time	Concerns regarding toxicity of the nanomaterial and may not be possible to regenerate the sensor	Medium	Pérez-López and Merkoçi, 2011

SOURCE: Alahi and Mukhopadhyay, 2017.

to produce assays with greater sensitivity and specificity. Noble metals (e.g., gold and silver) are frequently used for signal amplification because of their unique physicochemical properties; however, carbon, magnetic, metal oxide–based, and quantum dot nanoparticles have also been used (Rhouati et al., 2017). Incorporation of a nucleic acid amplification step into the biosensor design, particularly those that do not require temperature cycling (e.g., loop-mediated isothermal amplification, recombinase polymerase amplification, and rolling circle amplification) can also increase analytical sensitivity (Giuffrida and Spoto, 2017). Examples of nanosensors in developing specific food safety applications are detailed in Wang and Duncan (2017) and in Vigneshvar et al. (2016) (see some examples in Table 4-2).

TABLE 4-2 Selected Applications of Nanosensors in Food Safety

Type	Principle	Application	References[a]
Acetylcholinesterase inhibition–based biosensors	Electrochemistry	Understanding pesticidal impact	Pundir and Chauhan, 2012
Piezoelectric biosensors	Electrochemistry	Detecting organophosphate and carbamate	Marraza, 2014
Quartz-crystal biosensor	Electromagnetic	For developing ultra-high-sensitive detection of proteins and liquid	Ogi, 2013
Microbial fuel cell–based biosensors	Optical	To monitor biochemical oxygen demand and toxicity in the environment and heavy metal and pesticidal toxicity	Gutierrez et al., 2015; Sun et al., 2015
Based on cellulose nanocrystals	Nanomaterials	To detect norovirus	Rosilo et al., 2014
Based on aptamers, single-stranded oligonucleotides (DNA or RNA) that interact with analytes with antibody-like ability	Nanomaterials	To monitor mycotoxins in various foods (e.g., wine, ground corn)	Pak et al., 2014; Xiao et al., 2015

[a]From Vigneshvar et al., 2016.
SOURCES: Vigneshvar et al., 2016; Rhouati et al., 2017.

Sensor technologies are also highly applicable to monitoring product freshness, such as detecting biochemical parameters that are correlated with product spoilage and shelf life, particularly near product life end (Xiaobo et al., 2016). These types of sensors are usually noninvasive in nature. Examples of product attributes that can be measured are color, the presence of surface defects, and chemical composition. Technological platforms include optical, acoustical, NMR, and electrical. For example, light in the visible/near-infrared spectra can penetrate readily into biological systems, and when applied to a food can provide a "fingerprint" to assess parameters such as freshness, firmness, and texture. Biomimetic devices such as electronic noses, which are already used for personalized medicine (Fitzgerald et al., 2017), are being piloted for evaluating spoilage and shelf life of meats (Wojnowski et al., 2017).

At the end of the sensing phase, an electronic reader allows signal processing so that results are displayed in a user-friendly manner. Mobile diagnostics that use Internet-of-Things technologies to link sensor output to smartphones and cameras, and are even coupled with data entry on servers or the cloud, have been reported, particularly for detection of foodborne pathogens, food allergens, antibiotic residues, and shellfish toxins, in relevant sample matrices (Rateni et al., 2017). Although handheld mobile readouts are still in development with significant need for improvement (e.g., reducing signal-to-noise ratios, miniaturization, sample preparation, data interpretation, cost, and reliability), their future is bright because they provide options for portability and real-time results, important features for managing an already complex food chain.

3.2.1 Gaps

There are a number of practical impediments to successful, routine use of biosensor technologies in foods and environmental samples. Many different biosensors for detecting pathogens such as *Salmonella* in foods have been reported, but they vary widely in performance, particularly analytical sensitivity (detection limit), frequently ranging from a low of one single cell to a high of millions (Cinti et al., 2017; Silva et al., 2018). While of lesser importance in clinical settings where samples come from ill individuals and pathogen concentrations are high, this is not the case for food, water, and environmental samples. These sample types may be infrequently contaminated and when the contaminant is present, concentrations are low. Hence, for sensors to be of the greatest value in food safety, analytical sensitivity (detection limits) must be high (<10 cells) and specificity needs to be high. In addition, sample size should be large and testing done frequently in order to account for low contaminant prevalence.

In short, assay sensitivity and specificity (detection limits and low propensity for false positive and false negative results) will need to improve for sensor technologies to gain more widespread use in food and agriculture. In addition, there is a pressing need to develop sample preparation methods and protocols that will efficiently concentrate and purify an analyte from the matrix prior to use in the sensing device (Brehm-Stecher et al., 2009). This includes validating sensor performance in relevant natural sample matrices (i.e., various waters, foods, and environmental samples). Other factors related to conditions and ease of use, robustness, and cost are critical for success. Solving these practical scientific problems—along with ensuring that sensors are "fit for purpose"—will require extensive, transdisciplinary effort.

3.3 Opportunity 3: Integrated Data Management Systems

The development of omics and sensor technologies will augment the capabilities to collect increasing amounts of data in the food processing, safety, and quality realms (e.g., process validation, optimization and control, environmental, and public health data). Ultimately, the value of data to the food supply chain is to provide more and better information on which to base system optimization and management decisions regarding food processing, safety, quality control (e.g., preservation of product traits), food waste reduction, and system monitoring, among others. That means there must be an infrastructure to house massive amounts of records, and a means by which those records can be integrated and effectively used for decision-making purposes. Associated with these changes is the need to identify and understand the design and behaviors in emerging food supply chains.

In the area of food safety, there are a number of large, publicly accessible online databases used by the public health sector (inventoried in Marvin et al., 2017). Examples are the National Outbreak Reporting System, the Genome Trakr Network (whole-genome sequences of pathogens), and others used by industry, such as Combase (data for quantitative microbiology and models predict growth and inactivation of microorganisms). The availability of online searchable databases (e.g., JIFSAN's FoodRisk.org, a metadatabase for tools and models) and of social media and crowdsourcing (e.g., iwaspoisoned.com) platforms provide other capabilities. These and other databases have clear utility for different types of applications. However, that utility would be enhanced if data and databases were integrated with one another, particularly with less publicly accessible data such as industry process monitoring and product tracking systems (e.g., GPS and radio frequency identification [RFID]; Mejia et al., 2010), quality control systems, or public food safety monitoring efforts (see also Chapter 7 on data science). There are some early examples (see Box 4-4), but a much more concerted effort is needed toward data integration to

BOX 4-4
Early Examples of Data Integration

An example of successful data integration to help prioritize public response and health planning is the World Health Organization global platform for food safety data and information, FOSCOLLABO. This platform is designed to integrate data and information on food, hazard, country, and year through user interfaces or "dashboards" that can be customized by need and data format.

A quite different example is the Consortium for Managing the Food Supply Chain, a joint effort by IBM and Mars, Inc., that involves sequencing the microbiomes of foods along the supply chain to produce a genome sequence reference database for microbial pathogens. By comparing genome sequences from a test food to the reference, low levels of pathogens in an unknown sample can be identified very rapidly. With the proper analytics, this method could also be applicable to identification of product adulteration or the presence of specific traits for product verification.

SOURCES: Beans, 2017; WHO, 2018.

support food-safety decision making. Hill et al. (2017) provided a proof of concept for the use of genomic and epidemiological metadata integration along with sophisticated data analytics and modeling for early detection of human infection with foodborne pathogens. The model allowed this group to find associations between DNA sequence, location of the food animal across the production chain, and human illness. With technologies that collect, transmit, store, and analyze data obtained from real-time sensors, along with a centralized system of databases with sampling and processing data, their approach holds promise and would theoretically allow the tracking and tracing of individual food units.

Integrated data and data management systems can also be applied with the goal of improving resource efficiencies in the food system. Implementation of such a data management system in food networks supports efforts to optimize food processes and recycle and reduce waste during and after manufacturing as an operational concept in order to achieve a "circular economy" that makes best use of the range of waste streams in the agricultural and food system (see, for instance, the North American Initiative on Food Waste Reduction and Recovery [CEC, 2018] and the European Union's AgroCycle [UCD, 2018]). A more integrated and holistic data management systems approach to thinking about minimizing product and food waste may focus on identifying food waste conversion methods for edible and nonedible purposes. In such a distribution system, a seamless supply chain of foods and ingredients with a leaner, simpler, and transparent data management system would be vital. Considerations of system design and science are discussed in Chapter 8.

An example of an innovative exchange system developed to facilitate the diversion of surplus retail food products to distribution sites is the one recently developed with food banks. Efficiencies are gained through the development of networks and exchanges for distribution of surplus foods from retailers to food banks and distribution centers. Integration of real-time data on available foods and existing needs provides a mechanism for redirecting food to help feed the hungry and reduce food waste (Prendergast, 2017). Within the industry itself, goals for reducing food waste can be accomplished by setting standards for ordering, receiving, preparing, processing, packaging, serving, and tracking food production.

One technology that has enormous potential to revolutionize the management and storage of data and to facilitate the integration of food distribution systems, among other applications, is blockchain (see Chapter 7 on data science). Blockchain (also called open, distributed ledgers) is a system in which a continuously growing list of decentralized and encrypted records (blocks) are linked so that it can be securely distributed across peer-to-peer networks. Blockchain allows for highly transparent and instantaneous transfer of product data associated with many attributes, including the safety and quality of food, as well as environmental stewardship, all arising from activities such as routine monitoring, inspection and audit, accreditation, and laboratory analyses. The improvements from implementation of blockchain technology will also benefit consumers as they demand more detailed information about product sourcing, origin, processes, and production methods. For example, offering verified product sourcing, non-GM (genetically modified) or "organic" products to the consumer requires systems to preserve and track segregation through the supply chain. Consumers and other buyers are able to access information on the product via smartphone applications and other data platforms. Although in its infancy, blockchain is an important emerging technology (although it has its detractors) that may allow integration of detailed information across different platforms and ownership structures and provide verifiable information that consumers seek from manufacturers' claims. Although some applications that link purchases to specific retail outlets or consumers have benefits at the consumer level (e.g., in the case of food safety or other product recalls), little is known about consumer response to data systems extended to the retail consumer level. See Box 4-5 for more concrete examples of the application of blockchain technology to the management of data on the food supply chain.

BOX 4-5
Example 2: Making Risk Management Decisions by Using Blockchain and Associated Data Analytics (Recall, Outbreaks)

During an outbreak investigation or a product recall, a reliable traceability system is essential. In a comprehensive study, Mejia et al. (2010) reported that food companies have many different traceability methods, from as simple as manual records to as sophisticated as electronic based systems, although paper records are used by almost all parts of the supply chain to some degree (Mejia et al., 2010). Traceability is enforceable according to the 2011 Food Safety Modernization Act (P.L. 111-353), meaning that food establishments are required to keep detailed records but it is far from seamless. There is inconsistency in both the types of data collected and the way the data are recorded (i.e., from manual pen and paper to bar codes, RFID, and electronic systems). These all impact the speed at which information can be accessed and processed, an important consideration in the case of recalls and outbreaks. Although there have been initiatives to streamline product tracing, they are difficult to integrate into existing processes.

Walmart, Unilever, Dole, and others have recently collaborated with IBM to investigate potential roles for blockchain technology. Using the Hyperledger Fabric platform, companies are reporting that the increased transparency of blockchain would likely facilitate greater responsiveness in the case of foodborne illness outbreaks and recalls, including greater timeliness and accuracy. In fact, Walmart has already demonstrated the utility of blockchain to improve food safety by testing its efficacy in a mock trace-back of a fresh fruit product, which took 7 days using conventional methods but only 2.2 seconds by accessing blockchain records. Blockchain identified the farm from which the product was harvested and its path to the retail shelves.

3.3.1 Gaps

As companies are increasingly exploring the uses of blockchain technology in the agriculture and food arena, both challenges and solutions are arising. A recent report aimed at better understanding the implications and needs of the blockchain technology to stakeholders (e.g., producers, manufacturers, traders, and product standard organizations) identified the following key challenges: access and implementation of the technology, the need for a workforce that can adapt and learn new competencies, privacy concerns, considerations related to regulatory frameworks, interoperability, and compatibility with existing systems (Ge et al., 2017). Specific stakeholder groups have identified cost and knowledge of the technology as main challenges (IFIC Foundation, 2018).

3.4 Opportunity 4: Materials Sciences and Engineering

Food scientists apply engineering principles to design novel processing and packaging technologies that result in profound improvements to the quality, safety, acceptability, and shelf life of foods. Depending on the technology (e.g., thermal, aseptic, microwave, pulsed light, ohmic heating, high pressure, freezing, and refrigeration), these processes offer advantages such as improved product quality (organoleptic characteristics that resemble the fresh product), reduced energy usage, smaller footprint (better portability), and lower environmental impact (Neetoo and Chen, 2014). Some of these technologies may be particularly well suited for certain foods or venues, especially those in which large capital outlay for food processing is not possible or economically feasible.

Advances in materials science and nanotechnology, as applied to production of packaging materials, holds great promise for advancing quality and safety of food products. Active packaging, of which modified atmosphere packaging is an early example, is a system in which the food, the material, and the environment interact dynamically by incorporating oxygen scavengers, antimicrobials, and/or moisture adsorbents into the food packaging materials (Mlalila et al., 2016). These active compounds may or may not be released into the food, or prevent unwanted substances from entering the package, and, in so doing, they can improve product quality, ensure safety, and/or extend shelf life. For example, nanocomposite materials (i.e., polymers in combination with nanoparticles) provide both barrier and chemical protection to foods (Pradhan et al., 2015). "Smart" food packaging refers to a system that undergoes automatic changes in micro- or nanostructures as a consequence of dynamic changes to the environment (Mlalila et al., 2016). The materials that have smart properties are those able to control their interfacial properties. These largely consist of self-cleaning, self-cooling, and self-heating technologies, already designed for the health care sector, that are now being applied to food systems. Intelligent packaging systems are able to monitor the conditions, quality, and/or safety of a food, particularly during distribution and storage, and provide the consumer some evidence of product status (Mlalila et al., 2016). In some ways, this technology relates back to the biosensors discussed above in Opportunity 2. The output of intelligent packaging can be expressed in the form of data (e.g., at the level of specific product or lot using barcodes, RFID, or digital watermark) or as light (e.g., light-emitting diodes or holograms). All of these provide information that can then be included as a basis for decision-making and management systems. While monitoring food quality and freshness with indicators is routine in the food industry sector, intelligent packaging technologies are extremely well suited for detecting metabolites occurring as a consequence of food spoilage, and thus may have relevance for reducing food loss and waste at both the industry and consumer levels (see example in Box 4-6). From the consumer perspective, communicative packaging has emerged as a potential tool to address concerns about product quality, safety, and the consumer demand for specific product information as they make purchasing decisions.

3.4.1 Gaps

Although alternative food processing and packaging technologies have the potential to deliver better quality, nutrition, safety, and acceptability to food products, some questions related to the need to decrease the energetic footprint (e.g., energy and water savings, reliability) and environmental impacts (e.g., emissions or environmental degradation due to the use of plastic packaging materials) are unresolved. Likewise, acceptability of these

BOX 4-6
Example 3: Advanced Food Packaging Technologies

The combination of emerging technologies such as nanotechnology and materials science and engineering are propelling the production of advanced food packaging technologies. A long-recognized way to keep food fresh is to reduce oxygen exposure, which prevents microbial growth and enzymatic reactions, both of which contribute to food spoilage. Modified atmosphere packaging has been used for decades to reduce oxygen content in packages, but it is possible to make this technology "intelligent" by designing packages to include sensor technologies that measure oxygen levels. In 2007, a sensor such as this was used to specifically develop a food package. The ultraviolet (UV) light-activated sensor had an irreversible response and was reusable. The nanoparticulate crystalline titania was selected for its design because it had greater photoactivity. These materials were incorporated in a polymer to create an oxygen-sensitive, UV-activated film that could be printed directly onto food packaging material (Mills and Hazafy, 2008)—hence, intelligent packaging.

These early prototypes eventually led to a commercial product. In 2017, the smart indicator FRESHCODE was commercialized as an innovation to indicate degree of "freshness" of packaged chicken. The indicator contains an ink that captures the emission of volatile gases released during microbial growth and changes color as more gases are produced, indicative of a lower degree of product freshness. The visual readout can be accessed throughout the handling, storage, distribution, and sale of the product. It is particularly valuable for distribution chains, poultry product processors, and packers. In addition to serving as a freshness indicator for manufacturers, retailers, and consumers, this approach has the potential to reduce food waste by allowing for the timely diversion of the food product to food rescue and recovery channels.

relatively new technologies by the consumer still poses questions. Relative to nanomaterials, consideration of potential unintended consequences of their use is critical. Safety concerns focus on the potential interactions between nanomaterials and the food matrix, particularly potential toxicity to consumers and environmental impacts. Because these materials are very recent in their introduction to the market, there are relatively few data available to systematically assess health or environmental risks, and legislators err on the side of caution when it comes to regulatory decision making. Similarly, consumer acceptance of new technologies may be an issue and depend ultimately on the degree of trust consumers place on the products themselves (Roosen et al., 2015).

4. BARRIERS TO SUCCESS

4.1 Barrier 1: Consumer Acceptance

One important barrier to the implementation of technological advances in the food science and technology area is the need to better understand and anticipate consumers' food-related behaviors and choices, including the role of social and environmental factors, and underlying receptiveness to and understanding of information about products and processes. Identifying factors that determine consumer acceptance and choices over product attributes and qualities is essential information to determining the success in producing foods that will be purchased and consumed (e.g., Lusk et al., 2014).

Traditionally, consumers respond to market prices and other monetary signals in their product selection. However, there is increased evidence that financial incentives (such as taxes and subsidies applied to products), social factors, and context of food choices, as well as other behavioral motivators or nudges can encourage or discourage food-related behavior. Ignoring the need to better understand and anticipate consumer food behaviors, drivers, and trade-offs may limit consumer acceptance of new products, technologies, and market innovations. The need to better account for consumers' perceptions of risk around new technologies also underpins the need for education and strategies to best communicate the nature of food production, processing technologies, and the science involved so that consumers can make thoughtful and informed decisions in food selection, handling, and preparation. This applies to the need for effective food labeling approaches as well as basic communication about scientific and technological advances. A 2017 National Academies of Sciences, Engineering, and Medicine report (NASEM, 2017) highlights the need to understand the optimal communication approaches for use under different circumstances, and to recognize that many people do not make food selection and choice decisions based solely on scientific evidence.

4.2 Barrier 2: Regulatory Context

Scientific advancements in technologies related to food processing and product design, packaging, and handling may be limited by existing regulations, such as food law and product identity standards. A few examples are provided here. Many of the emerging food processing technologies (i.e., ohmic heating, ultrasound, or pulsed light) have not been validated for their ability to meet the mandated microbial inactivation standards for protection of public health. It may not be prudent from food safety and liability standpoints to use these processes commercially until such validations are conducted and reviewed. The inclusion of nanotechnology-based products (e.g., in packaging materials and for microencapsulation) may be met with regulatory scrutiny because these are not composed of materials that are generally recognized as safe. There is also the possibility that sensor devices or novel packaging materials may be prohibited based on current food adulteration regulations. The replacement of pathogen culture methods with whole-genome sequencing is being questioned because historically, proof of product adulteration in recall or outbreak situations relies on having a pure culture of the implicated organism, not simply evidence of the presence of its DNA. The practical use of technologies intended to collect data at a faster rate may be hindered if they have negative effects on other aspects of the process that fall under regulatory scrutiny, such as adhering to maximum line speeds in meat processing plants. Integrated and blockchain data systems offer the opportunity to digitize record keeping, some of which may be relevant for regulatory purposes (e.g., data from hazard analysis and critical control points or other preventive controls plans). However, relevant agencies may not yet be able to accommodate transfer of information using their current data management systems.

4.3 Barrier 3: Economics and Other Considerations

A relatively large share of investment in innovation and technologies for foods is done through the private sector where private returns to investment dictate technology choice with less emphasis placed on the public benefit. However, there remains a critical need for basic sciences and applications in which the payoffs advance science more broadly to benefit the public's and the private sector's interests. Furthermore, some basic research requires significant investment in underlying infrastructure. As an example, system-wide innovation and data networks often require large, upfront expenditures to develop and support data infrastructures.

However, interoperability of systems and data networks between the various participants in the supply chain is required to effectively monitor and maintain the safety and integrity of the food system, and to support efforts to integrate sustainability opportunities. With funding predominantly from private sources, the allocation of resources to research and research infrastructure may not address the highest-priority public needs.

Several of the scientific advances discussed above will provide more improved instrumentation and allow for collection of more sophisticated data. Training will be necessary to ensure that the existing and emerging workforce has the scientific skills to use these instruments, analyze the data, and make appropriate decisions that capitalize on the value of these new technologies.

Ultimately, consumer practices and food choice will determine the ability of product and process development to successfully improve product safety, quality, and design. Advances in behavioral sciences and effective communication about science, technology, risk, and decision-making communication are required to underpin successful adoption in the market.

5. RECOMMENDATIONS

Emerging technologies (e.g., omics, biosensors, and nanotechnology) have the potential to advance or transform the production of high-quality, safe, nutritious, and sustainable food products that meet the needs and demands of a diverse consumer population. Solving the fundamental and applied scientific problems necessary to use these technologies more widely will require multidisciplinary collaboration and funding mechanisms. Research efforts need to be transdisciplinary, involving not only food scientists but also those in other disciplines ranging from data and computer science, engineering, synthetic biology, and the social sciences, and many more. The committee identified the following high-priority research areas:

- Profile and/or alter food traits for desirability (such as chemical composition, nutritional value, intentional and unintentional contamination, and quality and sensory attributes) via improvements in processing and packaging technologies, the design and functionality of sensors, and the application of "foodomic" technologies.
- Provide enhanced product quality, nutrient retention, safety, and consumer appeal in a cost-effective and efficient manner that also reduces environmental impact and food waste by developing, optimizing, and validating advanced food processing and packaging technologies.
- Support improved decision making to maximize food integrity, quality, safety, and traceability, as well as to reduce food loss and

waste by capitalizing on data analytics, integration, and the development of advanced decision support tools.

- Enhance consumer understanding and acceptance of innovations in food production, processing, and safe handling through expanded knowledge about consumer behavior and risk-related decisions and practices.

REFERENCES

Alahi, M. E. E., and S. C. Mukhopadhyay. 2017. Detection methodologies for pathogens and toxins: A review. *Sensors* 17:1885.

Andjelkovic, U., M. S. Gajdosik, D. Gaso-Sokac, T. Martinovic, and D. Josic. 2017. Foodomics and food safety: Where we are. *Food Technology and Biotechnology* 55(3):290-307.

Beans, C. 2017. Inner workings: Companies seek food safety using a microbiome approach. *Proceedings of the National Academy of Sciences of the United States of America* 114(51):13306-13308.

Bellemare, M., M. Cakir, H. H. Peterson, L. Novak, and J. Rudi. 2017. On the measurement of food waste. *American Journal of Agricultural Economics* 95(5):1148-1158.

Brehm-Stecher, B., C. Young, L. Jaykus, and M. L. Tortorello. 2009. Sample preparation: The forgotten beginning. *Journal of Food Protection* 8:1774-1789.

Buzby, J. C., F. W. Hodan, and J. Hyman. 2014. *The Estimated Amount, Value, and Calories of Postharvest Food Losses at the Retail and Consumer Levels in the United States.* Economic Information Bulletin EIB-121, U.S. Department of Agriculture, Economic Research Service.

Capozzi, F., and A. Bordoni. 2013. Foodomics: A new comprehensive approach to food and nutrition. *Genes & Nutrition* 8:1-4.

CEC (Commission for Environmental Cooperation). 2018. *Characterization and Management of Food Waste in North America: Foundational Report.* Available at http://www.cec.org/islandora/en/item/11774-characterization-and-management-food-waste-in-north-america-foundational-report (accessed July 6, 2018).

Cinti, S., G. Volpe, S. Piermarini, E. Delibato, and G. Palleschi. 2017. Electrochemical biosensors for rapid detection of foodborne *Salmonella*: A critical overview. *Sensors* 17:1919.

Clancy, M., K. Fuglie, and P. Heisey. 2016. U.S. agricultural R&D in an era of falling public funding. *Amber Waves*, November 10. Available at https://www.ers.usda.gov/amber-waves/2016/november/us-agricultural-rd-in-an-era-of-falling-public-funding (accessed May 9, 2018).

Conrad, Z., M. T. Niles, D. A. Neher, E. D. Roy, N. E. Tichenor, and L. Jahns. 2018. Relationship between food waste, diet quality, and environmental sustainability. *PLoS ONE* 13(4):e0195405.

Corella, D., O. Coltell, G. Mattingley, J. V. Sorli, and J. M. Ordovás. 2017. Utilizing nutritional genomics to tailor diets for the prevention of cardiovascular disease: A guide for upcoming studies and implementations. *Expert Review of Molecular Diagnostics* 17(5):495-513.

Fitzgerald, J. E., E. T. H. Bui, N. M. Simon, and H. Fenniri. 2017. Artificial nose technology: Status and prospects in diagnostics. *Trends in Biotechnology* 35(1):33-41.

Floros, J. D., R. Newsome, W. Fisher, G. V. Barbosa-Cánovas, H. Chen, C. P. Dunne, J. B. German, R. L. Hall, D. R. Heldman, M. V. Karwe, and S. J. Knabel. 2010. Feeding the world today and tomorrow: The importance of food science and technology. *Comprehensive Reviews in Food Science and Food Safety* 9(5):572-599.

Gallo, M., and P. Ferranti. 2016. The evolution of analytical chemistry methods in foodomics. *Journal of Chromatography A* 1428:3-15.

Ge, L., C. Brewster, J. Spek, A. Smeenk, and J. Top. 2017. *Blockchain for Agriculture and Food: Findings from the Pilot Study.* Wageningen Economic Research Report 2017-112. Wageningen, The Netherlands: Wageningen University and Research Centre.

Giuffrida, M. C., and S. Spoto. 2017. Integration of isothermal amplification methods in microfluidic devices: Recent advances. *Biosensors and Bioelectronics* 90:174-186.

Grimaldi, K. A., B. van Ommen, J. M. Ordovas, L. D. Parnell, J. C. Mathers, I. Bendik, L. Brennan, C. Celis-Morales, E. Cirillo, H. Daniel, and B. de Kok, A. El-Sohemy, S. J. Fairweather-Tait, R. Fallaize, M. Fenech, L. R. Ferguson, E. R. Gibney, M. Gibney, I. M. F. Gjelstad, J. Kaput, A. S. Karlsen, S. Kolossa, J. Lovegrove, A. L. Macready, C. F. M. Marsaux, J. A. Martinez, F. Milagro, S. Navas-Carretero, H. M. Roche, W. H. M. Saris, I. Traczyk, H. van Kranen, L. Verschuren, F. Virgili, P. Weber, and J. Bouwman. 2017. Proposed guidelines to evaluate scientific validity and evidence for genotype-based dietary advice. *Genes & Nutrition* 12(1):35.

Gunders, D. 2012. *Wasted: How America Is Losing Up to 40 Percent of Its Food from Farm to Fork to Landfill.* NRDC Issue Paper IP:12-06-B. Washington, DC: Natural Resource Defense Council. Available at https://www.nrdc.org/sites/default/files/wasted-food-IP.pdf (accessed July 6, 2018).

Gutierrez, J. C., F. Amaro, and A. Martin-Gonzalez. 2015. Heavy metal whole-cell biosensors using eukaryotic microorganisms: An updated critical review. *Frontiers in Microbiology* 6:48.

Hill, A. A., M. Crotta, B. Wall, L. Good, S. J. O'Brien, and J. Guitian. 2017. Towards an integrated food safety surveillance system: A simulation study to explore the potential of combining genomic and epidemiological metadata. *Royal Society Open Science* 4:160721.

IFIC (International Food Information Council). 2018. *Blockhain: A Possible Breakthrough in Food Safety?* Available at https://www.foodinsight.org/blockchain-technology-foodsafety-foodwaste (accessed June 22, 2018).

Kaput, J., G. Perozzi, M. Radonjic, and F. Virgili. 2017. Propelling the paradigm shift from reductionism to systems nutrition. *Genes & Nutrition* 12:3.

Lee, S., D. O'Dell, J. Hohenstein, S. Colt, S. Mehta, and D. Erickson. 2016. NutriPhone: A mobile platform for low-cost point-of-care quantification of vitamin B_{12} concentrations. *Scientific Reports* 6:28237. doi: 10.1038/srep28237.

Lusk, J. L., and B. C. Briggeman. 2009. Food values. *American Journal of Agricultural Economics* 91(1):184-196.

Lusk, J. L., and J. McCluskey. 2018. Understanding the impacts of food consumer choice and food policy outcomes. *Applied Economic Perspectives and Policy* 40(1):5-21.

Lusk, J. L., J. Roosen, and A. Bieberstein. 2014. Consumer acceptance of new food technologies: Causes and roots of controversies. *Annual Review of Resource Economics* 6:381-405.

Mandal, P., A. Biswas, K. Choi, and U. Pal. 2011. Methods for rapid detection of foodborne pathogens: An overview. *American Journal of Food Technology* 6:87-102.

Marraza, G. 2014. Piezoelectric biosensors for organophosphate and carbamate pesticides: A review. *Biosensors* 4:301-317.

Marvin, H. J. P., E. M. Janssen, Y. Bouzembrak, P. J. M. Hendriksen, and M. Staats. 2017. Big data in food safety: An overview. *Critical Reviews in Food Science and Nutrition* 57(11):2286-2295.

Marwaha, S. S., and J. F. Kennedy. 1988. Whey—pollution problem and potential utilization. *International Journal of Food Science* 23(4):323-336.

Mejia, C., J. McEntire, K. Keener, M. K. Muth, W. Nganje, T. Stinson, and H. Jensen. 2010. Traceability (product tracing) in food systems: An IFT report submitted to the FDA, volume 2: cost considerations and implications. *Comprehensive Reviews in Food Science and Food Safety* 9(1):159-175.

Mills, A., and D. Hazafy. 2008. A solvent-based intelligence ink for oxygen. *Analyst* 133(2): 213-218.

Mlalila, N., D. M. Kadam, H. Swai, and A. Hilonga. 2016. Transformation of food packaging from passive to innovative via nanotechnology: Concepts and critiques. *Journal of Food Science and Technology* 53(9):3395-3407.

NASEM (National Academies of Sciences, Engineering, and Medicine). 2017. *Communicating Science Effectively: A Research Agenda*. Washington, DC: The National Academies Press.

NASEM. 2018. *Nutrigenomics and the Future of Nutrition: Proceedings of a Workshop—in Brief*. Washington, DC: The National Academies Press.

Neetoo, H., and H. Chen. 2014. Alternative food processing technologies. In *Food Processing: Principles and Applications*, 2nd ed., edited by S. Clark, S. Jung, and B. Lamsal. Hoboken, NJ: Wiley-Blackwell, pp. 137-169.

Neff, R. A., M. L. Spiker, and P. L., Truant. 2015. Wasted food: U.S. consumers' reported awareness, attitudes, and behaviors. *PLoS ONE* 10(6):0127881.

Ogi, H. 2013. Wireless-electrodeless quartz-crystal-microbalance biosensors for studying interactions among biomolecules: A review. *Proceedings of the Japan Academy, Series B, Physical and Biological Sciences* 89:401-417.

Pak, Y., S.-M. Kim, H. Jeong, C. G. Kang, J. S. Park, H. Song, R. Lee, N. Myoung, B. H. Lee, S. Seo, J. T. Kim, and G.-Y. Jung. 2014. Palladium-decorated hydrogen-gas sensors using periodically aligned graphene nanoribbons. *ACS Applied Materials & Interfaces* 6:13293-13298.

Pérez-López, B., and A. Merkoçi. 2011. Nanomaterials based biosensors for food analysis applications. *Trends in Food Science & Technology* 22:625-639.

Pradhan, N., S. Singh, N. Ojha, A. Shrivastava, A. Barla, V. Rai, and S. Bose. 2015. Facets of nanotechnology as seen in food processing, packaging, and preservation industry. *BioMed Research International*. Available at http://dx.doi.org/10.1155/2015/365672.

Prendergast, C. 2017. How food banks use markets to feed the poor. *Journal of Economic Perspectives* 31(4):145-162.

Pundir, C. S., and N. Chauhan. 2012. Acetylcholinesterase inhibition-based biosensors for pesticide determination: A review. *Analytical Biochemistry* 429:19-31.

Rateni, G., P. Dario, and F. Cavallo. 2017. Smartphone-based food diagnostic technologies: A review. *Sensors* 17:1453.

Rhouati, A., G. Bulbul, U. Latif, A. Hayat, Z.-H. Li, and J. L. Marty. 2017. Nano-aptasensing in mycotoxin analysis: Recent updates and progress. *Toxins* 9(11):349.

Roosen, J., A. Bieberstein, S. Blanchemanche, E. Goddard, S. Marette, and F. Vandermoere. 2015. Trust and willingness to pay for nanotechnology food. *Food Policy* 52:75-83.

Rosilo, H., J. R. McKee, E. Kontturi, T. Koho, V. P. Hytönen, O. Ikkala, and M. A. Kostiainen. 2014. Cationic polymer brush-modified cellulose nanocrystals for high-affinity virus binding. *Nanoscale* 6(20):11871-11881.

Silva, N. F. D., J. M. C. S. Magalhaes, C. Freire, and C. Delerue-Matos. 2018. Electrochemical biosensors for *Salmonella*: State of the art and challenges in food safety assessment. *Biosensors and Bioelectronics* 99:6678-6682.

Singh, T., S. Shukla, P. Kumar, V. Wahla, V. K., Bajpai, and I. A. Rather. 2017. Application of nanotechnology in food science: Perception and overview. *Frontiers in Microbiology* 8:1501.

Sun, J.-Z., G. P. Kingori, R.-W. Si, D.-D. Zhai, Z.-H. Liao, D.-Z. Sun, T. Zheng, Y.-C. Yong. 2015. Microbial fuel cell-based biosensors for environmental monitoring: A review. *Water Science & Technology* 71(6):801-809.

Toole, A. A., and F. Kuchler. 2015. *Improving Health Through Nutrition Research: An Overview of the U.S. Nutrition Research System.* Economic Research Report No. 182. U.S. Department of Agriculture Economic Research Service. Available at https://www.ers.usda.gov/webdocs/publications/45340/50982_err-182_report-summary.pdf?v=42030 (accessed May 4, 2018).

UCD (University College Dublin). 2018. AgroCycle. Available at http://www.agrocycle.eu (accessed June 22, 2018).

Vigneshvar, S., C. C. Sudhakumari, B. Senthilkumaran, and H. Prakash. 2016. Recent advances in biosensor technology for potential applications—an overview. *Frontiers in Bioengineering and Biotechnology* 4. doi: 10.3389/fbioe.2016.00011.

Wang, Y., and T. V. Duncan. 2017. Nanoscale sensors for assuring the safety of food products. *Current Opinion in Biotechnology* 44:74-86.

WHO (World Health Organization). 2016. *Biofortification of Staple Crops.* e-Library of Evidence for Nutrition Actions (eLENA). Available at http://www.who.int/elena/titles/biofortification/en (accessed July 6, 2018).

WHO. 2018. *FOSCOLLAB: Global Platform for Food Safety Data and Information.* Available at http://www.who.int/foodsafety/foscollab/en (accessed June 22, 2018).

Wojnowski, W., T. Majchrzak, T. Dymerski, and J. Gebicki. 2017. Electronic noses: Powerful tools in meat quality assessment. *Meat Science* 131:119-313.

Xiao, R., D. Wang, Z. Lin, B. Qiu, M. Liu, L. Guo, and G. Chen. 2015. Disassembly of gold nanoparticle dimers for colorimetric detection of ochratoxin A. *Analytical Methods* 7:842-845.

Xiaobo, Z., H. Xiaowei, and M. Povey. 2016. Non-invasive sensing for food reassurance. *Analyst* 141:1587-1610.

Zhang, G. 2013. Foodborne pathogenic bacteria detection: An evaluation of current and developing methods. *Meducator* 1(24):27-30. Available at https://journals.mcmaster.ca/meducator/issue/view/59.

5

Soils

1. INTRODUCTION

Fertile soils are among the greatest natural assets of the U.S. agricultural system. These include highly productive types of soils (e.g., mollisols and alfisols) that cover more than 30 percent of the country (USDA, 1999; see Figure 5-1). The importance of maintaining the nation's agricultural soils cannot be overstated: Soils are the basis of the nation's capacity to produce unparalleled crop yields and bountiful pasturelands, and, for all practical purposes, they are a finite resource. Soil formation is a slow, primarily geological process augmented by activity of biota living in soil, with formation rates in the productive U.S. heartland averaging a little more than one-tenth of a millimeter per year (Cruse et al., 2013). Many agricultural and other land-use practices cause physical, biological, and chemical degradation of soil that weakens its ability to support plant life and to provide ecosystem services. These conditions exacerbate the negative effects of weather and climate on soil stability and quality and contribute to water and air pollution (NRC, 2010; ITPS, 2015; NSTC, 2016). Future agricultural productivity to meet the needs of a growing world population can be possible only if the nation's fertile soils are maintained, which mandates protecting soils from erosion and degradation and efficiently managing vital crop nutrients.

The capacity for soils to support crop production and buffer the environment against perturbations such as climate change, drought, flood, and environmental pollution are a result of the ongoing, sophisticated interplay of their physical, chemical, and biological components. As agricultural production intensifies, an advanced understanding of these dynamics

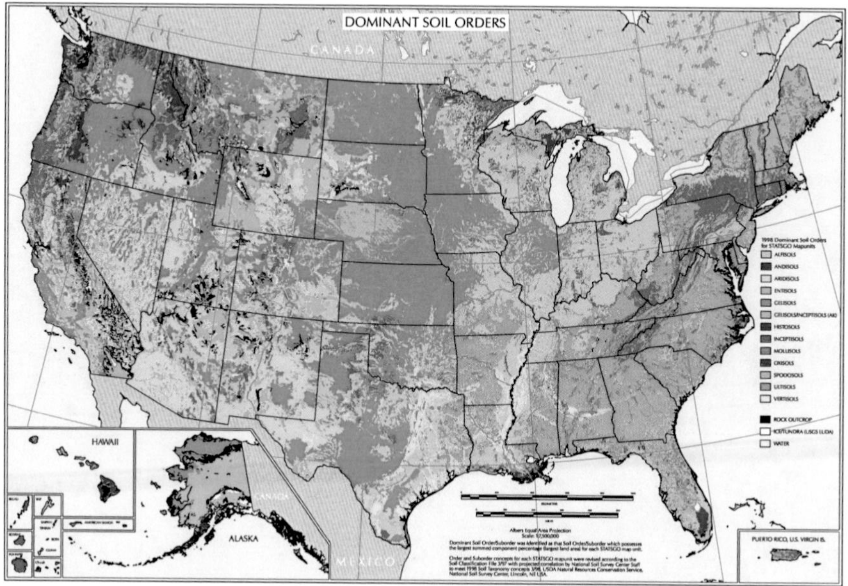

FIGURE 5-1 Dominant soil orders of the United States.
SOURCE: USDA NRCS, 2018.

under different management regimes and weather conditions is critical. This advanced understanding will require (1) the development and use of sensing technologies, biological strategies, and integrated systems approaches to maintain the depth and quality of existing fertile soils and restore degraded soils; (2) the combined use of novel sensing technologies, data analytics, precision plant breeding, and land management practices that significantly increase and optimize nutrient-use efficiency; and (3) supporting a healthy soil microbiome, which is instrumental to a wide range of soil ecosystem services, including nutrient production and bioavailability, breakdown of toxicants, and resilience to pests, pathogens, and other stressors such as climate variability and drought. This chapter describes a vision for a research strategy that will enable producers to sustain and enhance soil vitality for food production.

2. CHALLENGES

America's soils require stewardship. Threats to soil and consequently to food productivity fall into three broad categories: (1) soil sustainability, (2) soil quality, and (3) nutrient availability. The future for a vibrant, resilient

U.S. food production enterprise depends on understanding and taking action to address these threats.

2.1 Soil Sustainability

A primary challenge facing soil sustainability in the United States is erosion, a form of soil degradation that causes the displacement of the upper layers of soil by wind and water. Although soil erosion occurs under natural conditions, human activity on the land accelerates that process. Some agricultural practices, including conventional tillage, allowing cleared soil to remain uncovered, crop specialization (e.g., monoculture) with limited rotations or diversity, overuse of pesticides and fertilizers, and poor irrigation management, are associated with soil erosion and the breakdown of soil structure relative to the rate of soil formation (NRC, 2010; ITPS, 2015; NSTC, 2016). Extreme weather events and associated flooding further erode soil (Nearing et al., 2004).

Erosion not only displaces soil, but it also depletes soil organic matter (SOM) and reduces soil quality, fertility, and water-holding capacity (Magdoff and van Es, 2009) and thus reduces crop yields (den Biggelaar et al., 2003; Fenton et al., 2005). The results can be costly, with estimates ranging from hundreds of millions (Crosson, 2007) to tens of billions of dollars in the United States alone (Pimentel et al., 1995). Globally, erosion and the degradation of soil quality has led to the abandonment of approximately one-third of the world's arable land (UNCCD, 2017). Soil loss and erosion is not a new challenge, but has not yet been sufficiently addressed. Effective approaches to reduce soil erosion include no-till techniques; use of perennial, sod-forming crops; and use of cover crops (Montgomery, 2007; Magdoff and van Es, 2009). However, adoption of soil conservation practices are not widespread in the United States. In the United States alone, soil on cultivated cropland is estimated to be eroding at an average annual rate of ~2-5 tons/acre, with some regions losing soil at even greater rates, especially following severe weather events (Handelsman and Liautaud, 2016). However, the annual rate of soil formation is only on the order of 0.1 to 0.5 ton/acre. In the 1950s the United States developed guidelines for "tolerable" soil loss (Montgomery, 2007), but given the slow rate of soil formation, these guidelines with respect to long-term soil sustainability are under question. Without coordinated action, it has been posited that the United States is on track to run out of topsoil before the end of the 21st century (Handelsman and Liautaud, 2016).

2.2 Soil Health

The degradation of soil health[1] is caused by a wide range of interacting agricultural and nonagricultural factors. These include loss of nutrients and soil organic carbon through erosion and contamination through overuse of pesticides and fertilizers, increased salinization and acidification (NRC, 2010), and biologically by increases in soil-borne pathogens.

A key component of a healthy, fertile soil is soil organic carbon (SOC)—the major component of SOM—that serves as a key nutritional resource for plants and microbes. Although Earth's soils are a rich reservoir of carbon, with twice the amount of carbon than is held in the atmosphere and more than three times the amount of carbon held in vegetation, current agricultural practices have a predominantly detrimental impact on SOC. Intensive tillage practices cause declines of SOC stocks due to oxidation of organic matter, destruction of soil aggregates, and reduction in water infiltration (Olson et al., 2014a). Cultivation has been shown to significantly reduce SOC stocks by 30 perecent when tilling native prairie and forest soils of the north-central United States (Lal, 1999). Loss of SOC due to erosion is more pronounced on sloping land areas, due to drainage from the soil. SOC stocks declines are also predicted to be exacerbated by climate change with severe implications for future agricultural productivity (Weismeier et al., 2016).

Although agricultural expansion has resulted in modest increases of SOC in naturally infertile areas through practices such as liming, weed control, and fertilizer application (Spera et al., 2016), the potential to restore lost SOC is limited for most agricultural areas. Sanderman et al. (2017) estimated a global carbon debt due to agriculture of 116 petagrams of carbon for the top 2 m of soil, with the rate of loss increasing over the past 200 years. These losses are similar to carbon losses from vegetation due to deforestation. Hotspots for SOC loss are found in most major cropping regions and grazing lands around the world, including in the United States (see Figure 5-2).

The prevalence of diseases causing soilborne pathogens (seedling, vascular, and root rot diseases) is another important dimension of soil health. Disease contributes heavily to crop losses. The persistence of disease-causing pathogens is influenced by abiotic and biotic soil components and by agricultural practices (such as irrigation, tillage, and fertilization). Some management practices, such as crop rotation, have been shown to help avoid pathogen selection (Katan, 2017). In degraded soils or under continuous monoculture practices the beneficial members of the rhizosphere com-

[1]Also referred to as soil quality, soil health is the capacity of soils to function as living systems, with ecosystem and land use boundaries, to sustain plant and animal productivity, maintain or enhance water and air quality, and promote plant and animal health (FAO, 2008).

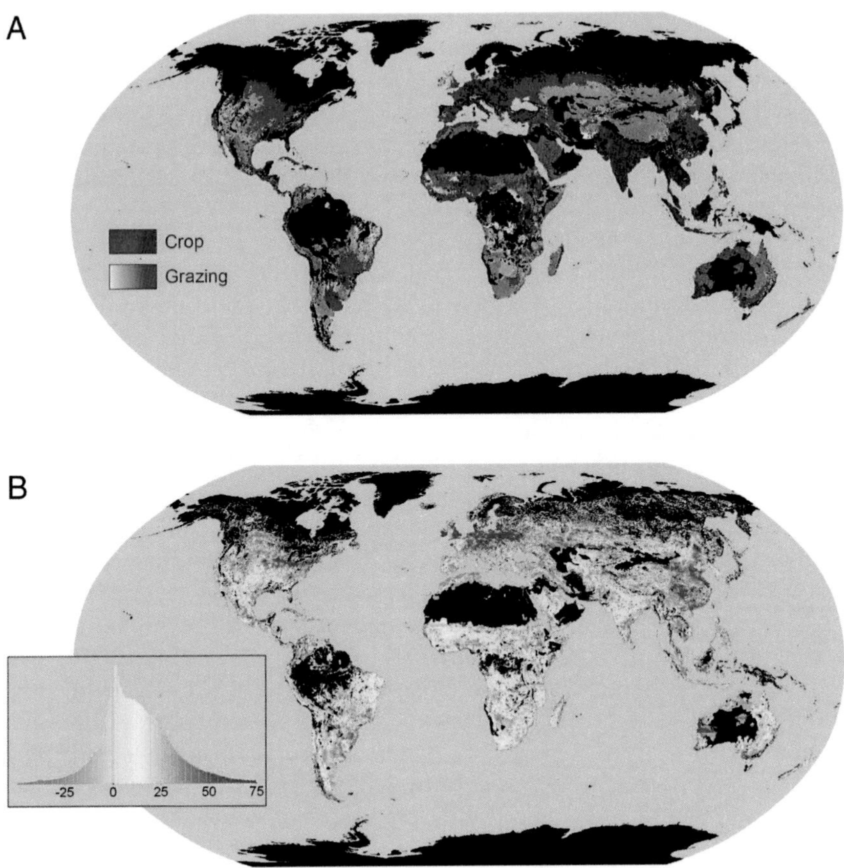

FIGURE 5-2 Global distribution of cropping and grazing in 2010. A. Color gradients indicate proportion of land use for crop production (red) and grazing (green). B. Histogram of SOC loss (Mg C ha^{-1}), with positive values indicating loss and negative values indicating gains in SOC.
SOURCE: Sanderman et al., 2017.

munity are sometimes unable to maintain a biological barrier to pathogens that thrive as a result (Lal, 2004; Busby et al., 2017).

2.3 Nutrient Availability

Nutrient availability in adequate amount, time, and place is key to success for any agricultural production system. Farmers rely heavily on chemical fertilizers to provide key plant macro- and micronutrients. Together with improved plant genetics, fertilizer use has generated enormous increases in crop productivity in recent decades. Unfortunately, continued dependence on chemical fertilizers to provide essential nutrients is unsustainable. The use of chemical fertilizers is causing well-documented negative consequences for the environment, including eutrophication, groundwater contamination, soil acidification, soil biodiversity loss, and the buildup of chemicals toxic to humans and animals (NRC, 2010).

The amount of nutrients naturally present in soils varies widely, and generally is insufficient to support high demands of crop production. For example, nitrogen, which is widely available in the atmosphere but more limited in soil, frequently becomes the first nutrient constraint on crop production (Lal and Stewart, 2017). The Haber-Bosch process has enabled the production of synthetic nitrogen fertilizer from atmospheric nitrogen (N_2) (Erisman et al., 2008). Overreliance and excessive application of synthetic nitrogen to ensure sufficient nitrogen supply for crops has, however, led to a significant reduction in global nitrogen-use efficiency (NUE),[2] from 68 percent in the 1960s to below 50 percent today (Lassaletta et al., 2014). Nitrogen is highly mobile in soils, and hence any amount applied that exceeds the crop uptake results in its loss. Too little nitrogen application leads to lower crop productivity, world hunger and malnutrition, and soil degradation (Sanchez and Swaminathan, 2005; Zhang et al., 2015; Lal and Stewart, 2017). Applying too much nitrogen results in environmental pollution and in the subsequent negative impacts on ecosystem health and human health (Jerrett et al., 2009; Avnery et al., 2011; Robertson et al., 2013; Zhang et al., 2015). Low NUE accounts for significant economic and environment loss, human health impacts, and a contribution of more than 67 percent of the N_2O emissions into the atmosphere. Many estimate that current levels of nitrogen fertilizer use have far exceeded the planetary boundaries (de Vries et al., 2013; Steffen et al., 2015), leading to increased calls for precision approaches to improve NUE.

Phosphorus is another essential nutrient for plant growth. However, the amount of phosphorus naturally present in soil varies. To compensate for deficiencies in bioavailable soil phosphorus, rock phosphate is normally applied to soil as fertilizer. There is growing concern among some experts that the worldwide stock of rock phosphate is rapidly depleting (Gilbert, 2009). As a result, advanced approaches to reuse phosphorus in waste products such as sewage sludge and manure will be critically important alongside the implementation of precision management practices that reduce the application of fertilizer. Rock phosphate is also a source of

[2]Nitrogen-use efficiency is defined as the ratio between the amount of fertilizer nitrogen removed from the field by the crop and the amount of fertilizer nitrogen applied (Saravia et al., 2016).

environmental contaminants such as cadmium, lead, chromium, nickel, manganese, and copper, which are toxic to soil, animal, and human health (Faridullah et al., 2017). The tendency to over-apply phosphorus (and other nutrients) to compensate for insufficient bioavailability of the individual nutrient components exacerbates negative environmental impacts. Phosphorus in fertilizer often binds with soil minerals so that it becomes unavailable to plants or is lost through runoff. Depending on land use and management practice, such runoff losses are as high as 42 percent (Hart et al., 2004), and upon reaching aquatic systems can result in environmental degradation due to eutrophication.

Continued reliance on chemical fertilizers as the dominant source of nutrients threatens the vitality of the soils on which agriculture relies and the long-term sustainability of the agricultural enterprise. Change will require new advances in identifying less harmful sources of nutrients, new approaches to improve the bioavailability of nutrients, and new mechanisms to recycle nutrients such as phosphorus from waste streams. These new sources and approaches are likely to enhance fertilizer use efficiency if accompanied by determining the "right" approaches of what, when, where, how much, and how fertilizers are applied.

3. OPPORTUNITIES

From the perspective of the sustainable uses of soil, the objective is to keep soil healthy, to be in places where it is needed, and to maximize its ability to provide essential goods and services, whether that is in crop production, water filtration, or carbon sequestration. An understanding of those processes, including the role of the soil microbiome in them, could help inform management of the status of the soil resource. Novel tools are emerging to probe, measure, monitor, and analyze soil and its characteristics in all of its complexity, from the molecular level to the ecosystem scale. Such tools, coupled with new approaches to farmer engagement and technology adoption, will enable both scientists and the farming community to sustainably maximize the benefits of soil resources.

3.1 Leveraging Advances in Microelectronics, Sensing, and Modeling

Soil ecosystems are marked by chemical, biological, and physical reactions that are affected by external perturbations such as weather or tillage, and the changes caused by these forces on soils have been documented over time. However, "visualizing" dynamic soil processes in real time has been an elusive goal in soil science, much less acquiring that information from a wide range of locations and under different environmental conditions and a diversity of management practices. Now, achieving that goal has the potential to become a reality, if the scientific community can facilitate the convergence of several fields of science and technology to build the tools that can enable this capability.

With progress in wireless communications and the miniaturization of electronic devices, coupled with advances in chemical, molecular, and optics signaling platforms, it is now possible to conceive of networks of sensor arrays positioned in situ on and below the soil surface across a farm field—potentially across many farm fields—that would actively measure the biological, chemical, and physical reactions taking place in the soil. These advances will provide the potential to acquire data about soils (erosion, moisture, nutrient content, biological activity, etc.) at the temporal and spatial resolutions that are needed to inform improved models of soil ecosystems and to enable implementation of site-specific management practices that reverse soil and nutrient loss. In the specific case of soil erosion, there is an opportunity to integrate high-resolution remote sensing, GPS, and GIS to survey soils and identify areas where soil losses are in need of specific conservation practices (Sahu et al., 2015). For example, fusion of high-resolution LISS III and PAN satellite imagery was used to make decisions about vegetative barriers to prevent soil erosion and to preserve soil moisture. Tiwari (2004) suggested bioengineering measures to reduce soil loss in areas of high to very high risk of soil erosion.

Application of sensor arrays, in situ and remote, would transmit data wirelessly to receiving devices above ground, connected to computers for collation and analysis of the data collected. Next-generation sensors, when fully developed and integrated into improved data decision models, will revolutionize our capability to deploy solutions in near real time and, in many cases, prior to the visible damage to plants (i.e., when they are asymptomatic). For example, georeferenced in situ soil sensors widely distributed across a farm may provide a measure of soil and plant nutrient status as a colored geospatial map, allowing farmers to site-specifically and variably apply nutrients, optimize inputs, and maximize output (Koch et al., 2004).

Having this kind of information would vastly expand fundamental knowledge about the drivers of soil dynamics. A better understanding of those dynamics in relation to crop production could give rise to actionable information for farmers, with the detail, speed, and affordability that currently do not exist. In the most expansive form envisioned, networks of arrays would produce high-resolution maps of biogeochemical activity—essentially, a dynamic "weather map" for the soils across the nation that would fill a missing gap in models of nutrient cycling, greenhouse gas production, and carbon storage.

3.2 Leveraging Advances in Soil Microbiology, Omics, Synthetic Biology, and Computational Biology to Harness Beneficial Properties of the Soil Microbiome

Soils have an incredible diversity of microbial life (Thompson et al., 2017), with billions of microbial cells and thousands to millions of taxa per gram (Torsvik et al., 2002; Knietch et al., 2003; Gans et al., 2005; Roesch et al., 2007; Jansson, 2011). Several soil microorganisms have previously been shown to improve plant growth through growth stimulation, nutrient provision, or disease suppression (Toyota and Watanabe, 2013; Bhardwaj et al., 2014; Courty et al., 2015; de Souza et al., 2015; Mus et al., 2016; Baez-Rogelio et al., 2017; Gupta et al., 2017). However, the vast majority of soil microbes and their potential beneficial functions remain to be discovered. Therefore, an opportunity is to gain a better understanding of this rich resource of soil microbial biodiversity and to exploit beneficial functions of the soil microbiome to not only optimize crop productivity, but also to provide other beneficial ecosystem services, such as sequestration of organic carbon and increasing water retention.

Greater understanding of the diversity and importance of microbes individually and collectively is now emerging, owing in part to the research field and tools of metagenomics[3] and associated omics technologies (Biteen et al., 2016). These omics data have the potential to be leveraged to inform the development and design of new microbial products, including combinations of microorganisms (consortia) that might be more resilient to environmental stress, as components of targeted biofertilizer and biocontrol formulations. A specific example is to exploit the ability of arbuscular mycorrhizal fungi and associated bacteria to mobilize and increase the bioavailability of soil phosphorus to plants (Hinsinger et al., 2015). Another opportunity is to use synthetic biology approaches for custom design of rhizosphere microbiomes that are optimally suited to enhance productivity of a specific crop under a given soil type and climate regime (Busby et al., 2017). In addition, by exploration of the omics data generated from soil microbiomes, there is the opportunity to use this knowledge to exploit and custom-design microbial species and enzymes for new processes, such as new pathways for affordable and sustainable nitrogen fertilizer production (Liu et al., 2017).

Finally, research is currently under way to elucidate the "metaphenome" of the soil microbiome, that is, the functions carried out, or phenotypes, of the microbiome, as a totality of the genetic expression of the collective microorganisms, in response to environmental conditions and

[3]Metagenomics is a field of science that involves genome-level characterization and understanding of microbes and their communities, through high-throughput studies that incorporate methods from genomic and other omics studies (NRC, 2007).

available resources (Jansson and Hofmockel, 2018). By examining how the physiology, metabolism, and interactions of the soil microbiome change in response to a perturbation, such as moisture, investigators hope to develop predictive models of the microbiomes' behavior and state. This research will be an important dovetail with other new powerful methods of measuring the collective environmental response of the soil at higher scales.

In summary, this opportunity will build on the convergence of soil microbiology, metagenomics and other omics, synthetic biology, and computational biology, together with advances in more traditional disciplines, including microbial physiology, fermentation, and formulation of seed/soil inoculants.

The committee notes the recent publication of opportunities for microbiome research across different agencies in the United States that is much in line with our own recommendations, *Interagency Strategic Plan for Microbiome Research, FY2018-2022* (MIWG, 2018). This plan included a strategy for manipulating the soil/plant microbiome to increase crop productivity, with specific recommendations to advance knowledge of the microbiome, including determining their functions and how they contribute to soil sustainability. It is too early to judge the success of this unified strategy; however, the committee supports the call for increased funding in microbiome science to support these potentially important advances in knowledge.

3.3 Integrating Social Sciences on Technology Adoption and Farmer Engagement into Soil Health Research

Research on soil loss and sustainability is fairly advanced. Some of the existing transformative tools and practices that better protect soils include reduced, modified, or no-tillage methods (also called "conservation tillage"), crop rotations, use of cover crops, split application of fertilizer, drip-line irrigation, and controlled or slow release of nutrients. Most of these methods not only prevent or reduce soil loss and degradation, but also have other environmental benefits such as carbon sequestration (Bernoux et al., 2006; Lal et al., 2007; Olson et al., 2014b; Poeplau and Don, 2015) and economic benefits such as reducing costs of labor and production (Allmaras and Dowdy, 1985; Mitchel et al., 2009; Vitale et al., 2011). Pairing known methods to reduce soil loss with site-specific soil sensors and models has the potential to become a powerful breakthrough approach to addressing erosion. However, U.S. farmers have not been successful in adopting some, much less all, of these soil conservation approaches (Mitchell et al., 2007; Bossange et al., 2016; Carlisle, 2016; Ulrich-Schad et al., 2017). The development of near-term technology and best management practices will require convergence of agronomy and social science research on technology adoption and farmer engagement.

A variety of factors influence the adoption of new technology and approaches to land management specifically regarding soils (e.g., reduced-till strategies and higher precision fertilization practices). Research suggests that the most prominent barriers include economics (affordability of new tools, labor, or change in methods), education (availability and accessibility of learning resources), social networks (community support networks to facilitate technology adoption), and farmer perceptions about the benefits of adopting new approaches. Carlisle (2016) provides a comprehensive review of research on each of these barriers and emphasizes that their influence on farmer decision making varies according to geographic, economic, and social contexts. A missing element, however, is research on engagement—involving farmers in development of tools and methods that reduce soil loss and promote soil health, and in the development of education and policy incentives to support technology adoption. Research on effective engagement strategies for the introduction of new technologies is rapidly expanding (Kleinman et al., 2007; Nisbet, 2009; Philbrick and Barandiaran, 2009; Sclove, 2010; Rask and Worthington, 2015). Bringing the science of engagement into research on and development of strategies that encourage technology adoption opens new opportunities for addressing the suite of challenges associated with soil loss and degradation.

4. GAPS

We currently lack the essential knowledge required to manage agricultural soil ecosystems to obtain their full potential. Recognition that soil's geophysical dimensions are highly influenced by its living components has brought to the forefront a need for better understanding of the interplay of soil's biological attributes with its physical and chemical components. Major knowledge gaps to be filled include the following.

4.1 Influence of Cropping Systems and Amendments on Soil Properties

Sufficient understanding of the physical and chemical properties of soil at appropriate temporal and spatial scales to model, manage, and predict changes in agricultural soil ecosystems is needed. In addition, better information is needed on how different cropping systems and amendments, such as manure, influence the soil texture, water storage capacity, resistance to erosion, and SOM retention. We also need to understand the interactions between soil physical, chemical, and biological properties and the ability of the soil microbiome to facilitate soil aggregation and soil carbon retention.

4.2 Interactions Among Soil Ecosystems, Nutrient Bioavailability, and Crop Productivity

SOM pools can provide a significant amount of nutritional needs, (particularly mineral nitrogen needs) of the crop throughout the growing season. However, there are no reliable techniques to quantify and account for nitrogen supplied to the crop from mineralization of SOM. A better understanding is needed of nutrient distribution in space, time, and depth and how it contributes to crop nutritional needs, particularly mineral nitrogen needs throughout the growing season. Better information on the role of the microbiome with respect to nutrient cycling—including increasing nitrogen-use efficiency, increasing bioavailability of soil phosphorus, and cycling of SOM—is needed. In addition, a better molecular-level understanding is needed of soil biogeochemistry and the influence of specific enzymes, metabolites, and signaling molecules on crop productivity.

4.3 Interactions Between Soil Ecosystems and Climate

Climate change will impact soil organic carbon flux to the atmosphere and cycling of other nutrients. Although these functions are vitally important for plant growth and sustainability of life on our planet, the mechanistic details are largely unknown. Currently, studies of bulk processes involved in organic carbon turnover in soil, such as decomposition and respiration, reflect the sum of myriads of specific metabolic processes carried out by interacting members of the soil community. An understanding is needed of the biogeochemical pathways involved in SOC decomposition and greenhouse gas emission, and how changes in environmental conditions influence those processes. Ultimately, this knowledge can lead to better management practices for sequestering soil SOC and preventing carbon loss from the system.

5. BARRIERS TO SUCCESS

5.1 Insufficient Tools and Techniques to Measure, Model, and Manage Soil Ecosystem Services Carried Out by the Soil Microbiome

Because of the high physical heterogeneity and biological diversity of soils, it has been a significant challenge to obtain relevant measures of ecosystem services carried out by the soil microbiome. Current technologies are largely based on high-throughput sequencing or measures of bulk processes (such as soil respiration) and are insufficient to decode the functional relevance of the soil microbiome at relevant scales (Biteen et al., 2016; Blaser et al., 2016). Although metagenomics datasets provide a bounty of phylogenetic and functional information, deciphering the roles of microbial genes and gene products has proven to be difficult due to the high microbial diversity in soil (Blaser et al., 2016; Jansson and Hofmockel, 2018). Also, it is challenging to generalize findings from one soil system to another because of the differences in soil characteristics, parent materials, climate, and other factors that shape soils. Thus, to fully understand, predict, and harness beneficial microbes for agricultural purposes, more research is needed across several different science disciplines, including soil microbiology and microbial ecology, combined with computational biology and advanced omics and imaging technologies. This multidisciplinary approach is necessary to elucidate the molecular mechanisms underpinning soil–microbe–plant interactions and to determine how to optimize beneficial interactions to improve agricultural production.

5.2 Limited Development and Deployment of Modern Sensor Technologies to Assess Soil Properties Across Different Scales of Resolution

Sensors have been deployed for decades to measure parameters of interest in agriculture, for example, sensing soil moisture to decide the time to irrigate crops or monitoring weather for crop planting. Many of these sensing methodologies, however, provide only a point measure. Recent advances in precision agriculture have brought innovative tools and implements to farmers, enabling them to manage farm inputs at a much finer scale than ever before. Currently, farmers' ability to precision-manage is limited by their capability to precision-measure. There is a growing need to measure multiple soil and plant properties at a very high frequency and data density, in heterogeneous space and time. This requires development of novel active and passive sensors that provide diagnostic measures of soil and crop properties of interest rapidly, reliably, nondestructively, in situ, and in motion, and that provide near-real-time information to farmers and scientists to increase efficiency of farming practices.

However, the development and deployment of advanced materials that can provide soil data at the molecular level or nanoscale, including biological sensors, is still in its infancy. Sensors are also needed, but have not yet been developed, that (a) can be buried indefinitely in the subsurface of soil, (b) can capture chemical signals that enable "imaging" of the rhizosphere, and (c) can wirelessly transmit data through the soil substrate to aboveground receivers. Depending on which process or reaction for which a measurement is sought, a chemical-biological-optical-electrical signaling platform must be devised. The sensors would need to be reliable, miniaturized, energy efficient, and of low cost, enabling them to be arranged in arrays. Such sensors might include plants themselves connected to microelectronic detection systems potentially coupled with remote sensing devices. It may also include the development and application of live microbial biosensors, for example, that are engineered using synthetic biology tools to produce a recordable signal in response to specific soil nutrient levels and/or needs of the growing plant (Xu et al., 2013; Lindemann et al., 2017; Wang et al., 2017).

5.3 Lack of Soil Data Commons to Share Data, Analytical Methods, and Models

There is a need to integrate and share soil physical, chemical, and biological data to harness the intellectual capital of scientists from multiple disciplines (from microbiology to data analytics) to translate data into meaningful knowledge. A common infrastructure with open access could be similar to the National Cancer Institute's Genomic Data Commons[4] or the Human Cell Atlas Data Coordination Platform.[5] A centralized community-driven platform for depositing and sharing data does not yet exist for the soil sciences, and this holds the field back. In addition, translating data into meaningful knowledge and, ultimately, management decisions requires the development of computational tools that integrate models of crop nitrogen demand with soil nitrogen supply to inform decision support tools that are the basis of precision agriculture. Such advances are necessary to realize the complete potential of the Five-R approach for precision agriculture—having the right input, at the right time, at the right place, in the right amount, and in the right manner (Khosla, 2010).

6. RECOMMENDATIONS FOR NEXT STEPS

Emerging science—including sensor engineering, data science, and microbiology—shows considerable promise for developing advanced tools to better assess, monitor, and enhance agricultural soils. The transformation from knowledge to action will require sustained investments in research and its associated infrastructure, such as unified data sharing and analysis platforms. Also, research efforts will need to be conducted through transdisciplinary collaborations that involve not only soil scientists, but also engineers, microbiologists, ecologists, and data scientists, among others. The committee identified the following high-priority goals that, with adequate support for research, have the potential to be achieved within the next decade:

---

[4]See https://gdc.cancer.gov.

[5]See https://www.humancellatlas.org/data-sharing.

- Maintain depth and health of existing fertile soils and restore degraded soils through adoption of best agronomic practices, combined with the use of new sensing technologies, biological strategies, and integrated systems approaches.
- Significantly increase and optimize nutrient-use efficiency (especially nitrogen) through the integration of novel sensing technologies, data analytics, precision plant breeding, and land management practices.
- Create more productive and sustainable crop production systems by identifying and harnessing the soil microbiome's capability to produce nutrients, increase nutrient bioavailability, and improve plant resilience to environmental stress and disease.
- Improve the transfer of technology and practices to farmers to reduce soil loss through converging research in soil sciences, technology adoption, and community engagement.

REFERENCES

Allmaras, R. R., and R. H. Dowdy. 1985. Conservation tillage systems and their adoption in the United States. *Soil & Tillage Research* 5:197-222.

Avnery, S., D. L. Mauzerall, J. Liu, and L. W. Horowitz. 2011. Global crop yield reductions due to surface ozone exposure: 2. Year 2030 potential crop production losses and economic damage under two scenarios of O_3 pollution. *Atmospheric Environment* 45(13):2297-2309.

Baez-Rogelio, A., Y. E. Morales-García, V. Quintero-Hernández, and J. Muñoz-Rojas. 2017. Next generation of microbial inoculants for agriculture and bioremediation. *Microbial Biotechnology* 10(1):19-21.

Bernoux, M., C. C. Cerri, C. E. P. Cerri, M. S. Neto, A. Metay, A.-S. Perrin, E. Scopel, R. Tantely, D. Blavet, M. C. de Piccolo, M. Pavei, and E. Milne. 2006. Cropping systems, carbon sequestration and erosion in Brazil, a review. *Agronomy for Sustainable Development* 26:1-8.

Bhardwaj, D., M. W. Ansari, R. K. Sahoo, and N. Tuteja. 2014. Biofertilizers function as key player in sustainable agriculture by improving soil fertility, plant tolerance and crop productivity. *Microbial Cell Factories* 13(1):66.

Biteen, J. S., P. C. Blainey, Z. G. Cardon, M. Chun, G. M. Church, P. C. Dorrestein, S. E. Fraser, J. A. Gilbert, J. K. Jansson, R. Knight, J. F. Miller, A. Ozcan, K. A. Prather, S. R. Quake, E. G. Ruby, P. A. Silver, S. Taha, G. van den Engh, P. S. Weiss, G. C. L. Wong, A. T. Wright, and T. D. Young. 2016. Tools for the microbiome: Nano and beyond. *ACS Nano* 10(1):6-37.

Blaser, M. J., Z. G. Cardon, M. K. Cho, J. L. Dangl, T. J. Donohue, J. L. Green, R. Knight, M. E. Maxon, T. R. Northen, K. S. Pollard, and E. L. Brodie. 2016. Toward a predictive understanding of Earth's microbiomes to address 21st century challenges. *mBio* 7(3):e00714-16.

Bossange, A. V., K. M. Knudson, A. Shresth, R. Harben, and J. P. Mitchell. 2016. The potential for conservation tillage adoption in the San Juaquin Valley, California: A qualitative study of farmer perspectives and opportunities for extension. *PLoS ONE* 11(12):e0167612.

Busby, P. E., C. Soman, M. R. Wagner, M. L. Friesen, J. Kremer, A. Bennett, M. Morsy, J. A. Eisen, J. E. Leach, and J. L. Dangl. 2017. Research priorities for harnessing plant microbiomes in sustainable agriculture. *PLoS Biology* 15(3):e2001793.

Carlisle, L. 2016. Factors influencing farmer adoption of soil health practices in the United States: A narrative review. *Agroecology and Sustainable Food Systems* 40(6):583-613.

Courty, P. E., P. Smith, S. Koegel, D. Redecker, and D. Wipf. 2015. Inorganic nitrogen uptake and transport in beneficial plant root-microbe interactions. *Critical Reviews in Plant Sciences* 34(1-3):4-16.

Crosson, P. 2007. Soil quality and agricultural development. Pp. 2911-2932 in *Handbook of Agricultural Economics, Volume 3: Farmers, Farm Production and Farm Markets,* edited by R. Evenson and P. Pingali. Amsterdam and New York: North Holland.

Cruse, R. M., S. Lee, T. E. Fenton, E. Wang, and J. Laflen. 2013. Soil renewal and sustainability. In *Principles of Sustainable Soil Management in Agroecosystems*, edited by R. Lal and B. A. Stewart. Boca Raton, FL: CRC Press.

de Souza, R., A. Ambrosini, and L. M. P. Passaglia. 2015. Plant growth-promoting bacteria as inoculants in agricultural soils. *Genetics and Molecular Biology* 38(4):401-419.

de Vries, F. T., E. Thébault, M. Liiri, K. Birkhofer, M. A. Tsiafouli, L. Bjørnlund, H. B. Jørgensen, M. V. Brady, S. Christensen, P. C. de Ruiter, T. d'Hertefeldt, J. Frouz, K. Hedlund, L. Hemerik, W. H. G. Hol, S. Hotes, S. R. Mortimer, H. Setälä, S. P. Sgardelis, K. Uteseny, W. H. van der Putten, V. Wolters, and R. D. Bardgett. 2013. Soil food web properties explain ecosystem services across European land use systems. *Proceedings of the National Academy of Sciences of the United States of America* 110(35):14296-14301.

den Biggelaar, C., R. Lal, K. Wiebe, and V. Breneman. 2003. The global impact of soil erosion on productivity, I: Absolute and relative erosion-induced yield losses. *Advances in Agronomy* 81:1-48.

Erisman, J. W., M. A. Sutton, J. Galloway, Z. Klimont, and W. Winiwarter. 2008. How a century of ammonia synthesis changed the world. *Nature Geoscience* 1(10):636.

FAO (Food and Agriculture Organization). 2008. An international technical workshop Investing in Sustainable Crop Intensification: The Case for Improving Soil Health. Integrated Crop Management. Rome, Italy. Available at http://www.fao.org/docrep/012/i0951e/i0951e00.htm (accessed July 6, 2018).

Faridullah, F., M. Umar, A. Alam, M. Amjad Sabir, and D. Khan. 2017. Assessment of heavy metals concentration in phosphate rock deposits, Hazara basin, Lesser Himalaya Pakistan. *Geosciences Journal* 21(5):743-752.

Fenton, T. E., M. Kazemi, and M. A. Lauterbach-Barrett. 2005. Erosional impact on organic matter content and productivity of selected Iowa soils. *Soil and Tillage Research* 81(2):163-171.

Gans, J., M. Wolinsky, and J. Dunbar. 2005. Computational improvements reveal great bacterial diversity and high metal toxicity in soil. *Science* 309(5739):1387-1390.

Gilbert, N. 2009. The disappearing nutrient. *Nature* 461:716-718.

Gupta, S., R. Kaushal, and G. Sood. 2017. Impact of plant growth–promoting rhizobacteria on vegetable crop production. *International Journal of Vegetable Science* 24(3):289-300.

Handelsman, J., and P. Liautaud. 2016. A call to action to save one of America's most important natural resources. Blog, White House Office of Science and Technology Policy. Available at https://obamawhitehouse.archives.gov/blog/2016/08/01/call-action-save-one-americas-most-important-natural-resources (accessed July 6, 2018).

Hart, M. R., B. F. Quin, and M. Nguyen. 2004. Phosphorus runoff from agricultural land and direct fertilizer effects. *Journal of Environmental Quality* 33(6):1954-1972.

Hinsinger, P., L. Herrmann, D. Lesueur, A. Robin, J. Trap, K. Waithaisong, and C. Plassard. 2015. Impact of roots, microorganisms and microfauna on the fate of soil phosphorus in the rhizosphere. Pp. 375-407 in *Annual Plant Reviews, Volume 48: Phosphorus Metabolism in Plants*, edited by W. C. Plaxton and H. Lambers. New York: John Wiley & Sons.

ITPS (Intergovernmental Technical Panel on Soils). 2015. *Status of the World's Soil Resources: Technical Summary.* Rome, Italy: FAO. Available at http://www.fao.org/documents/card/en/c/39bc9f2b-7493-4ab6-b024-feeaf49d4d01 (accessed June 12, 2018).

Jansson, J. 2011. Towards "tera terra": Terabase sequencing of terrestrial metagenomics. *Microbe* 6(7):309-315.

Jansson, J. K., and K. S. Hofmockel. 2018. The soil microbiome—from metagenomics to metaphenomics. *Current Opinion in Microbiology* 43:162-168.

Jerrett, M., R. T. Burnett, C. A. Pope III, K. Ito, G. Thurston, D. Krewski, Y. Shi, E. Calle, and M. Thun. 2009. Long-term ozone exposure and mortality. *New England Journal of Medicine* 360:1085-1095.

Katan, J. 2017. Diseases caused by soilborne pathogens: Biology, management and challenges. *Journal of Plant Pathology* 99(2):305-315.

Khosla, R. 2010. Precision agriculture: Challenges and opportunities in a flat world. 19th World Congress of Soil Science, Soil Solutions for a Changing World. August 1-6, 2010, Brisbane, Australia. Available at https://www.iuss.org/19th%20WCSS/Symposium/pdf/0779.pdf (accessed May 28, 2018).

Kleinman, D. L., M. Powell, J. Grice, J. Adrian, and C. Lobes. 2007. A toolkit for democratizing science and technology policy: The practical mechanics of organizing a consensus conference. *Bulletin of Science, Technology & Society* 27(2):154-169.

Knietsch, A., T. Waschkowitz, S. Bowien, A. Henne, and R. Daniel. 2003. Metagenomes of complex microbial consortia derived from different soils as sources for novel genes conferring formation of carbonyls from short-chain polyols on *Escherichia coli*. *Journal of Molecular Microbiology and Biotechnology* 5(1):46-56.

Koch, B., R. Khosla, D. G. Westfall, M. Frasier, and D. Inman. 2004. Economic feasibility of variable rate N application in irrigated corn. *Agronomy Journal* 96:1572-1580.

Lal, R. 1999. Soil management and restoration for C sequestration to mitigate the accelerated greenhouse effect. *Progress in Environmental Science* 1(4):307-326.

Lal, R. 2004. Soil carbon sequestration impacts on global climate change and food security. *Science* 304(5677):1623-1627.

Lal, R., and B. A. Stewart. 2017. *Urban Soils.* Boca Raton, FL: CRC Press.

Lal, R., D. Reicosky, and J. Hanson. 2007. Evolution of the plow over 10,000 years and the rationale for no-till farming. *Soil and Tillage Research* 93:1-12.

Lassaletta, L., G. Billen, B. Grizzetti, J. Anglade, and J. Garnier. 2014. 50 year trends in nitrogen use efficiency of world cropping systems: The relationship between yield and nitrogen input to cropland. *Environmental Research Letters* 9(10):105011.

Lindemann, S. R., J. M. Mobberley, J. K. Cole, L. M. Markillie, R. C. Taylor, E. Huang, W. B. Chrisler, H. S. Wiley, M. S. Lipton, W. C. Nelson, J. K. Fredrickson, and M. F. Romine. 2017. Predicting species-resolved macronutrient acquisition during succession in a model phototrophic biofilm using an integrated 'omics approach. *Frontiers in Microbiology* 8:1020.

Liu, C., K. K. Sakimoto, B. C. Colón, P. A. Silver, and D. G. Nocera. 2017. Ambient nitrogen reduction cycle using a hybrid inorganic–biological system. *Proceedings of the National Academy of Sciences of the United States of America* 114(25):6450-6455.

Magdoff, F., and H. van Es. 2009. *Building Soils for Better Crops.* Beltsville, MD: Sustainable Agriculture Network.

Mitchell, J. P., K. Klonsky, A. Shrestha, R. Fry, A. Dusault, J. Beyer, and R. Harben. 2007. Adoption of conservation tillage in California: Current status and future perspectives. *Australian Journal of Experimental Agriculture* 47:1383-1388.

Mitchell, J. P., K. M. Klonsky, E. M. Miyao, and K. J. Hembree. 2009. Conservation tillage tomato production in California's San Joaquin Valley. UC ANR Communication Services Publication 8330.

MIWG (Microbiome Interagency Working Group). 2018. *Interagency Strategic Plan for Microbiome Research FY 2018-2022*. Available at https://science.energy.gov/~/media/ber/pdf/workshop%20reports/Interagency_Microbiome_Strategic_Plan_FY2018-2022.pdf (accessed July 6, 2018).

Montgomery, D. R. 2007. Soil erosion and agricultural sustainability. *Proceedings of the National Academy of Sciences of the United States of America* 104(33):13268-13272.

Mus, F., M. B. Crook, K. Garcia, A. Garcia Costas, B. A. Geddes, E. D. Kouri, P. Paramasivan, M.-H. Ryu, G. E. D. Oldroyd, P. S. Poole, M. K. Udvardi, C. A. Voigt, J.-M. Ané, and J. W. Peters. 2016. Symbiotic nitrogen fixation and the challenges to its extension to nonlegumes. *Applied and Environmental Microbiology* 82(13):3698-3710.

Nearing, M. A., F. F. Pruski, and M. R. O'Neal. 2004. Expected climate change impacts on soil erosion rates: A review. *Journal of Soil and Water Conservation* 59(1):43-50.

Nisbet, M. C. 2009. Framing science: A new paradigm in public engagement. Pp. 40-67 in *Understanding Science: New Agendas in Science Communication*, edited by L. Kahlor and P. Stout. New York: Taylor & Francis.

NRC (National Research Council). 2007. *The New Science of Metagenomics: Revealing the Secrets of Our Microbial Planet*. Washington, DC: The National Academies Press.

NRC. 2010. *Toward Sustainable Agricultural Systems in the 21st Century*. Washington, DC: The National Academies Press.

NSTC (National Science and Technology Council). 2016. *The State and Future of U.S. Soils: Framework for a Federal Strategic Plan for Soil Science*. Available at https://obamawhitehouse.archives.gov/sites/default/files/microsites/ostp/ssiwg_framework_december_2016.pdf (accessed July 6, 2018).

Olson, K., S. A. Ebelhar, and J. M. Lang. 2014a. Long-term effects of cover crops on crop yields, soil organic carbon stocks and sequestration. *Open Journal of Soil Science* 4:284-292.

Olson, K. R., M. Al-Kaisi, R. Lal, and B. Lowery. 2014b. Examining the paired comparison method approach for determining soil organic carbon sequestration rates. *Journal of Soil and Water Conservation* 69(6):193A-197A.

Philbrick, M., and J. Barandiaran. 2009. The National Citizens' Technology Forum: Lessons for the future. *Science and Public Policy* 36(5):335-347.

Pimentel, D., C. Harvey, P. Resosudarmo, K. Sinclair, D. Kurz, M. McNair, S. Crist, L. Shpritz, L. Fitton, R. Saffouri, and R. Blair. 1995. Environmental and economic costs of soil erosion and conservation benefits. *Science* 267(5201):1117-1123.

Poeplau, C., and A. Don. 2017. Carbon sequestration in agricultural soils via cultivation of cover crops—a meta-analysis. *Agriculture, Ecosystems & Environment* 200(1):33-41.

Rask, M., and R. Worthington. 2015. *Governing Biodiversity through Democratic Deliberation*. New York: Routledge.

Robertson, G. P., T. W. Bruulsema, R. J. Gehl, D. Kanter, D. L. Mauzerall, C. A. Rotz, and C. O. Williams. 2013. Nitrogen–climate interactions in US agriculture. *Biogeochemistry* 114(1-3):41-70.

Roesch, L. F. W., R. R. Fulthorpe, A. Riva, G. Casella, A. K. M. Hadwin, A. D. Kent, S. H. Daroub, F. A. O. Camargo, W. G. Farmerie, and E. W. Triplett. 2007. Pyrosequencing enumerates and contrasts soil microbial diversity. *ISME Journal* 1(4):283-290.

Sahu, N., G. O. Reddy, N. Kumar, and M. S. S. Nagaraju. 2015. High resolution remote sensing, GPS and GIS in soil resource mapping and characterization: A review. *Agricultural Reviews* 36(1):14-25.

Sanchez, P. A., and M. S. Swaminathan. 2005. Hunger in Africa: The link between unhealthy people and unhealthy soils. *Lancet* 365(9457):442-444.

Sanderman, J., T. Hengl, and G. J. Fiske. 2017. Soil carbon debt of 12,000 years of human land use. *Proceedings of the National Academy of Sciences of the United States of America* 114(36):9575-9580.

Saravia, D., E. R. Farfán-Vignolo, R. Gutiérrez, F. De Mendiburu, R. Schafleitner, M. Bonierbale, and M. A. Khan. 2016. Yield and physiological response of potatoes indicate different strategies to cope with drought stress and nitrogen fertilization. *American Journal of Potato Research* 93(3):288-295.

Sclove, R. E. 2010. Perspectives: Reinventing technology assessment. *Issues in Science and Technology* 27(1).

Spera, S. A., G. L. Galford, M. T. Coe, M. N. Macedo, and J. F. Mustard. 2016. Land-use change affects water recycling in Brazil's last agricultural frontier. *Global Change Biology* 22(10):3405-3413.

Steffen, W., K. Richardson, J. Rockström, S. E. Cornell, I. Fetzer, E. M. Bennett, R. Biggs, S. R. Carpenter, W. de Vries, C. A. de Wit, C. Folke, D. Gerten, J. Heinke, G. M. Mace, L. M. Persson, V. Ramanathan, B. Reyers, and S. Sörlin. 2015. Planetary boundaries: Guiding human development on a changing planet. *Science* 347(6223):1259855.

Thompson, L. R., J. G. Sanders, D. McDonald, A. Amir, J. Ladau, K. J. Locey, R. J. Prill, A. Tripathi, S. M. Gibbons, G. Ackermann, J. A. Navas-Molina, S. Janssen, E. Kopylova, Y. Vázquez-Baeza, A. González, J. T. Morton, S. Mirarab, Z. Z. Xu, L. Jiang, M. F. Haroon, J. Kanbar, Q. Zhu, S. J. Song, T. Kosciolek, N. A. Bokulich, J. Lefler, C. J. Brislawn, G. Humphrey, S. M. Owens, J. Hapton-Marcell, D. Berg-Lyons, V. McKenzie, N. Fierer, J. A. Fuhrman, A. Clauset, R. L. Stevens, A. Shade, K. S. Pollard, K. D. Goodwin, J. K. Jansson, J. A. Gilbert, R. Knight, and the Earth Microbiome Project Consortium. 2017. A communal catalogue reveals Earth's multiscale microbial diversity. *Nature* 551(7681):457-463.

Tiwari, A. K. 2004. Effective bio-engineering measures for torrent and erosion control in Shivaliks. In *Compendium National Symposium on Enhancing Productivity and Sustainability in Hill and Mountain Agro-Ecosystem*, edited by V. N. Sharda. Dehradun, India: CSWCRTI.

Torsvik, V., L. Ovreas, and T. F. Thingstad. 2002. Prokaryotic diversity–magnitude, dynamics, and controlling factors. *Science* 296(5570):1064-1066.

Toyota, K., and T. Watanabe. 2013. Recent trends in microbial inoculants in agriculture. *Microbes and Environments* 28(4):403-404.

Ulrich-Scad, J. D., S. Garcia de Jalon, N. Babin, A. Pape, and L. S. Prokopy. 2017. Measuring and understanding agricultural producers' adoption of nutrient best management practices. *Journal of Soil and Water Conservation* 72(5):506-518.

UNCCD (United Nations Convention to Combat Desertification). 2017. *Global Land Outlook*, 1st ed. Bonn, Germany: Secretariat of the United Nations Convention to Combat Desertification. Available at https://global-land-outlook.squarespace.com/the-outlook/#the-bokk (accessed July 6, 2018).

USDA (U.S. Department of Agriculture). 1999. *Soil Taxonomy: A Basic System of Soil Classification for Making and Interpreting Soil Surveys*, 2nd ed. Washington, DC: USDA. Available at https://www.nrcs.usda.gov/Internet/FSE_DOCUMENTS/nrcs142p2_051232.pdf (accessed July 6, 2018).

USDA NRCS (Natural Resources Conservation Service). 2005. Global Soil Regions Map. Available at https://www.nrcs.usda.gov/wps/portal/nrcs/detail/soils/use/?cid=nrcs142p2_054013 (accessed July 6, 2018).

USDA NRCS. 2018. Distribution Maps of Dominant Soil Orders. Available at https://www.nrcs.usda.gov/Internet/FSE_MEDIA/stelprdb1237749.pdf (accessed November 21, 2018).

Vitale, J. D., C. Godsey, J. Edwards, and R. Taylor. 2011. The adoption of conservation tillage practices in Oklahoma: Findings from a producer survey. *Journal of Soil and Water Conservation* 66(4):250-264.

Wang, P., R. Khoshravesh, S. Karki, R. Tapia, C. P. Balahadia, A. Bandyopadhyay, W. Paul Quick, R. Furbank, T. L. Sage, and J. A. Langdale. 2017. Re-creation of a key step in the evolutionary switch from C_3 to C_4 leaf anatomy. *Current Biology* 27(21):3278-3287.

Wiesmeier, M., C. Poeplau, C. A Sierra, H. Maier, H., C. Frühauf, R. Hübner, A. Kühnel, P. Spörlein, U. Geuß, E. Hangen, and B. Schilling. 2016. Projected loss of soil organic carbon in temperate agricultural soils in the 21st century: Effects of climate change and carbon input trends. *Scientific Reports* 6:32525.

Xu, X., P. E. Thornton, and W. M. Post. 2014. A global analysis of soil microbial biomass carbon, nitrogen and phosphorus in terrestrial ecosystems. *Global Ecology and Biogeography* 22:737-749.

Zhang, X., E. A. Davidson, D. L. Mauzerall, T. D. Searchinger, P. Dumas, and Y. Shen. 2015. Managing nitrogen for sustainable development. *Nature* 528(7580):51-59.

6

Water-Use Efficiency and Productivity

1. INTRODUCTION

Freshwater is an essential input for agriculture that uses significant quantities of the U.S. water supply. It is projected that by 2050, average farm yields will need to double in major cereal systems in order to meet expected increases in food demand (Tilman et al., 2011). Freshwater is a finite resource; thus, the necessary increases in crop agricultural productivity (yields) can only be met with a significant increase in water-use efficiency. In addition, there are competing interests for water, including energy production, domestic and industrial needs, recreation, and maintaining environmental quality. In combination with population growth and increasing extreme weather events, these factors have already resulted in significant changes for U.S. agricultural water use. Some arid regions (e.g., west Texas) have reached a tipping point with low aquifer storage unable to meet agricultural water demands, forcing the use of high-salinity groundwater (Uddameri and Reible, 2018). Additional long-tail risks to the agricultural water supply could come from energy extraction activities (such as hydraulic fracturing) or carbon sequestration activities, which may contaminate subsurface freshwater supplies and make them unusable for agriculture without costly treatment (Vengosh et al., 2014). Sustainable intensification of agriculture and the associated need for sufficient freshwater to produce food will require a shift in water sources, treatment, use, reuse, and management.

This chapter describes the challenges in ensuring the availability of freshwater and optimizing the efficiency of water use in agricultural settings along with the scientific opportunities and gaps to overcome the challenges. There are tremendous near-term opportunities to improve water-use efficiency[1] and water productivity[2] through new technologies and systems-level approaches. The opportunities include (1) better use of spatial-resolution data and data science, (2) improving plant and soil properties to increase water-use efficiency, and (3) optimizing water use and reuse through systems-level management approaches and implementation of controlled environments. Improving water-use efficiency will require simultaneously applying various water-saving and water-optimizing approaches. Finally, the chapter identifies research and societal barriers that may impede progress in increasing water-use efficiency. As with the rest of this report, policies and regulations that might be needed to promote the use of these water-saving approaches are not addressed.

2. CHALLENGES

Crop production is water intensive, and crop productivity is dependent on the availability of water. Crop agriculture constitutes an estimated 80 percent of national consumptive water use in the United States (USDA-ERS, 2018), largely through irrigation. In 2012, irrigated farms accounted for approximately half of the total value of crop sales on 28 percent of U.S. harvested cropland (USDA-ERS, 2018). In 2010, total irrigation withdrawals averaged 115 billion gallons per day, whereas total water withdrawals for direct use in livestock production and aquaculture averaged only 2 and 10 billion gallons per day, respectively (USGS, 2016). Water stress is the largest contributor to U.S. crop loss, and low water availability affects approximately 45 percent of U.S. land surfaces (DeLucia et al., 2014). Continued access to freshwater for irrigated high-value agriculture will be critical for meeting future food demands.

While irrigation practices have improved over the past several decades—with shifts from flood to spray to pivot to drip irrigation that incrementally decrease water consumption—water productivity and water-use efficiency are well below what is achievable. Further decreases in water use to promote more sustainable crop agriculture will require a combination of revolutionary new approaches, such as second-generation drip irrigation combined with sensors and data analytics, improved and regionally based weather and seasonal climate forecasts, plants engineered to be more water

[1]Water-use efficiency is defined in this report in the hydrological context; it is the ratio of the volume of water used productively (Stanhill, 1986). This is the percentage of water supplied to the plant that is effectively taken up by the plant, that is, for example, not lost to drainage or bare soil evaporation.

[2]Water productivity is the yield in production per unit of water used.

efficient, smarter soils, and alternative sources of water that are accessible. This section describes important obstacles that need to be overcome in the coming decade.

2.1 Agricultural Productivity Is Dependent on Freshwater

For many regions in the United States, the current methods of agricultural water use are unsustainable. For example, groundwater aquifers store rainwater for the future, but in some locations and during periods of prolonged drought, groundwater is extracted at a faster rate than it is recharged. This has caused substantial decreases in the groundwater levels in the Central Valley of California and in the Kansas High Plains (Ogallala) aquifer in the Midwest. This has also resulted in greater pumping costs and increasingly saline waters for agriculture. The Ogallala aquifer is responsible for over 90 percent of irrigation water in the Central High Plains. Many regions of the aquifer are already depleted, and from 1960 to 2010, about 30 percent of the storage had already been consumed. It is estimated that at current use rates, as much as 80 percent of the stored water will have been consumed by 2060 (Steward et al., 2013; see also Figure 6-1). The use of water-intensive crops in regions where water is scarce (e.g., almonds and alfalfa in California), and shifts from snow-fed water systems to rain-fed water systems as the climate warms (Pederson et al., 2011), will also stress water-poor regions. This will limit agricultural productivity in the region if the area must revert back to reliance on only rain-fed agriculture. Advances in data science, sensing, modeling, water-use efficiency, and systems-level management practices can be applied to achieve sustainable water use that meet the needs of agriculture.

2.2 Spatial, Temporal, and Climate Variability

A major challenge for better water management is planning and preparedness for the high levels of spatial and temporal variability of conditions that affect water-use efficiency, such as climate variability. Man-made climate change is projected to increase the intensity of storms as well as the number and duration of dry spells, increasing uncertainty in water availability at weather timescales. Though subseasonal to seasonal forecasts are becoming more skillful, current seasonal-scale forecasts are still inaccurate (~60 percent certainty) and this can negatively impact yields (Brown and Lall, 2006). This is in part due to limited predictability of some of the phenomena underlying subseasonal to seasonal forecast predictability, such as natural modes of variability and elements of external forcing. There is also limited publicly available, spatially resolved data on water use for irrigated agriculture, such as total amounts pumped to and applied to crops. This

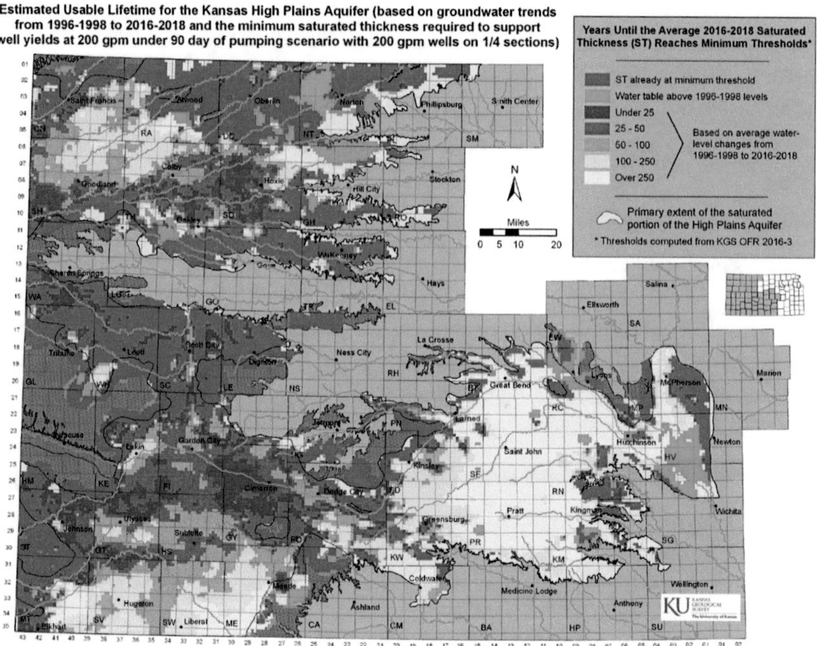

FIGURE 6-1 Map of the Kansas High Plains aquifer and groundwater levels.
NOTES: The map shows estimates for when groundwater levels in the Kansas High Plains (Ogallala) aquifer will be at minimum thresholds for use for different groundwater management districts (GMDs). Brown areas are already below minimum thresholds, while red areas have less than 25 years to reach those thresholds. Groundwater mining of these regions is a result of withdrawals being greater than recharge from precipitation.
SOURCE: Kisekka and Aguilar, 2016.

hinders understanding of water use and prevents making informed decisions about improving water-use efficiency. Greater reliability of forecast can help improve water management decisions, and improve operating efficiency of crops (Block and Goddard, 2012). In addition to precise use of water in agriculture, spatially resolved data on vegetative health, soil properties, nutrient concentrations, water quality parameters, and water quantity are needed to construct accurate prediction models for water-use efficiency. The data will need to be collected at a higher spatial and temporal resolution than is currently collected, especially for remote sensing methods such as satellite data.

3. OPPORTUNITIES

There are a number of research opportunities that could be transformative for reducing water use in agriculture, with many opportunities based on increasing water-use efficiency in crops. However, in some systems such as controlled environment agriculture (CEA), there is the potential to simultaneously increase both water-use efficiency and water productivity. Water use is fundamentally a challenge that cuts across many components of the food system, and the opportunities below are considered in the context of other chapters in this report, for example, in developing drought-, saline-, and flood-tolerant crops; engineering soil to improve its water-holding capacity; or creating animal housing facilities with higher water-use efficiency. The following opportunities have the greatest near-term potential to improve water-use efficiency in crops: (1) prescriptive analytics for maximizing water-use efficiency, (2) improved resiliency of plants with respect to water use and soils with respect to water-holding capacity, and (3) optimized water use and reuse through alternative water sources and CEA.

3.1 Prescriptive Analytics

Prescriptive analytics leverage an understanding of system behavior to prescribe (or proactively determine) the changes needed to operate optimally. The ability to cost-effectively monitor water demand and manage water supply at high spatial and temporal resolutions presents an opportunity to improve water-use efficiency for crop agriculture. To accomplish this will require ubiquitous data collection and seamless integration of data types and analytics across vastly different temporal and spatial scales (e.g., the Global Agriculture Monitoring project [CGLAMR, 2018]). Sensors on unmanned aerial vehicles that can move and measure at different locations, on-ground sensors, and satellite data can also be used together to provide higher resolutions of multiple parameters. The development of scalable and cost-effective technologies (e.g., sensors) would improve water-use management by allowing for more precise monitoring of moisture content at high spatial resolution over vast scales. These technologies could be combined with improved remote sensing capabilities at higher spatial resolutions to monitor for soil moisture, groundwater levels, or plant health. Water pricing and policy will need to provide the incentives required to promote adoption of these water-saving technologies (Olen et al., 2016).

Improved data collection, management, and sharing about water availability (quantity and quality), water demand, and water use at high spatial and temporal scales will be important for improving water-use efficiency in agriculture. These data can support ecohydrology-based approaches to water management. For example, improved seasonal climate forecasts can be used to predict future water quantities and the viability of various cropping systems. As the number of plants with different traits increases (see Chapter 2), that could also enable farmers to select the most climate-appropriate varieties of those crops (such as drought-tolerant cultivars). In addition, enhanced prediction of extreme and disruptive events as well as inclusion of more components of the Earth system in subseasonal to seasonal forecast models will better enable water use and management (NASEM, 2016).

A better understanding of the interactions between agroecosystems, natural ecosystems, and the built environment can help to improve water-use efficiency and water productivity. The ability to collect on-ground sensor data at high spatial resolution provides an opportunity to better understand the coupled nature of agroecosystems and natural ecosystems, and to refine physics-based (mechanistic) models describing system behaviors. Understanding the role of sociological and economic factors in water use for agriculture has the potential to further improve water management practices using integrated systems approaches (Stuart et al., 2015). This ability will help to limit unintended consequences to the environment, human activities, and the economy that may result from reducing water use.

3.2 Soil and Plant Resiliency

Water-use efficiency can be improved through better soil management practices and by engineering soil properties to maximize water availability to plants. Promising opportunities include manipulating the soil microbiome to improve water use by plants (especially under drought or salt stress conditions), increasing soil carbon through additions of engineered materials, and improving soil management (e.g., no till). There are also opportunities to improve water-use efficiency through real-time monitoring of plant health.

The role of the soil and plant microbiomes in affecting water availability to plants is largely unexplored. However, this ability could be exploited through development of new approaches to controlling evaporation from soils (a portion of water consumption from evapotranspiration) while enhancing water uptake by plants, particularly during periods of salt or drought stress (Marasco et al., 2012). Any improvements to promote water availability to plants at higher soil tension would increase water efficiency in soils. However, these advances cannot diminish other important functions of soils (see Chapter 5): storing organic carbon and other nutrients, filtering contaminants and pathogens, provision of nutrients, biodiversity habitat, and hydrological buffer.

Increased soil organic carbon increases the water-holding capacity of soils, particularly at the wilting point (Karhu et al., 2011; Minasny and

McBratney, 2017). Cost-effective methods to add natural or engineered carbon to soil coupled with best practices to manage water retention in soils (cover crops and no-till agriculture) provide a method to improve water-use efficiency (Turner, 2004). Better understanding of the optimal properties and sources of carbon to add to soils could lead to better engineered soils with respect to water-use efficiency (e.g., moisture-controlled water availability).

Spatially resolved real-time monitoring of plant health may also improve precision watering. Plant nanotechnology can provide methods to better monitor plant health status (Giraldo et al., 2014). Embedding nanomaterials in sentinel plants provides a signal that can be monitored remotely, potentially allowing the plants to communicate their health status with respect to water and nutrient availability. These data along with soil moisture content and other measurements will help to deliver water only where it is required.

3.3 Controlled Environments and Alternative Water Sources

Water-use efficiency can be improved by decentralizing parts of food production (e.g., fruits and vegetables) and better coupling it with the built environment. CEA offers systems-level opportunities to increase the sustainability of agriculture by providing resource-efficient farming systems with respect to water and nutrient use. For example, vertical farming of lettuce in controlled environments uses approximately 10 percent of water compared to traditional crop production in Yuma, Arizona (Barbosa et al., 2015). Fish production in recirculating aquaculture systems or aquaponics systems is highly efficient with respect to water use and enables reuse of nitrogen (waste) from fish production for plant production. These different types of CEAs also provide consistent year-round yields, fewer challenges with pests and disease, and the potential for high nutrient-use efficiency (Bregnballe, 2015; Harwood, 2017). These systems are currently used for producing fish protein as well as high-value crops such as vegetables, fruits, and herbs. Improvements in sensor technology, robotics, and artificial intelligence to monitor nutrient levels and plant health make controlled environments feasible and attractive. The integration of CEA within the built environment offers an additional opportunity to recover resources (e.g., water or nitrogen) from domestic or industrial sources, making this alternative more sustainable.

4. GAPS

4.1 Lack of Accurate Methods and Forecasts

A better understanding of the flows of water in the agroecosystem at high spatial and temporal resolution can be used to develop more accurate models for predicting and improving water-use efficiency. Deploying sensor networks will require development of small, biodegradable, inexpensive, energy-efficient disposable sensors. They need to provide depth-resolved soil moisture or nutrient levels to allow for more precise irrigation and nutrient applications. In planta sensors are needed to monitor plant health status with respect to water requirements, potentially at the scale of single plants.

Data on crops, soils, water, climate, and more are collected at multiple spatial and temporal scales using different techniques. Different types of data are stored in different databases, and even for a particular variable may be reported in different units by different sources. Better methods are needed to harmonize disparate datasets, to cross-reference data collected at different spatial scales, and to verify data and model veracity.

Data-driven analytics for Internet-of-Things–enabled water use-efficiency decision support systems will benefit from efficient cloud computing and local analysis due to the difficulty in mobilizing and storing high volumes of data. Automated methods for quality control of data and improved communications between sensors to activate them only when required will be essential. For example, a sensor upstream could communicate with a sensor downstream to prompt it to begin to collect data at a higher rate when spikes of salinity or nutrients are expected (McGuinness et al., 2014). "Smarter" sensors to reduce the data collection and processing demands and methods to distribute the processing of data across all of sensor platforms (e.g., Edge Computing) will also be needed for real-time monitoring and responses to needs or perturbations in the system.

Several research questions need to be addressed:

1. How can we develop and deploy sensor networks at high spatial and temporal scales to optimize water use at the system level?
2. How can we integrate ground-based and remote sensing data to best manage water use?
3. How can we improve computing speeds, data analytics, and data sharing to provide weather forecasts at high spatial and temporal resolution (<100 m)?
4. How can we improve relevance and thus use of seasonal climate forecasts?

4.2 Lack of Understanding of Plant and Soil Property Impacts on Water-Use Efficiency

There is a growing body of literature indicating that the soil microbiome changes in response to water availability (e.g., Manzoni et al., 2012), yet, there has been less emphasis on identifying how to manipulate the soil microbiome to enhance water availability and water-use efficiency. Significant improvement in water-use efficiency can also be achieved by controlling the stomatal opening in plants through genetic modification or using plant growth regulators (Moshelion et al., 2015; Glowacka et al., 2018). However, a limited understanding of the genes and plant physiology that control water uptake and transpiration, as well as drought and salinity resistance, hinders rapid selection of plants that are more water efficient, enabling the use of lower-quality water for agriculture. Improved understanding of how to control the rate of stomatal closing (e.g., in periods of shading) could be used to decrease transpiration without altering biomass production, food production, or nutritional content. Systems biology approaches to modeling water flows in agroecosystems may provide new approaches to water management.

While increasing soil carbon has the potential to improve water-use efficiency by lowering evaporation or drainage losses, the magnitude of the increases in water-use efficiency that this may provide is under debate. The mechanisms behind the effects of carbon on water availability to plants and water-use efficiency require more careful assessment (Minasny and McBratney, 2017).

Plant sensor technologies have great potential to monitor water-use efficiency. However, advances will require the development of appropriate devices (e.g., disposable biodegradable sensors) and practical methods to emplace nanomaterials into plants. The impacts of these sensors on plant health and food quality will also need to be determined.

Several research questions remain to be addressed:

1. What are the roles of the soil microbiome and plant traits in controlling water-use efficiency?
2. What are efficient methods to increase soil carbon content to improve water-holding capacity of soils?
3. How can plant and soil sensors be used together with weather forecasts, plant selection, and other technologies to optimize water use?

4.3 Need for Alternative Water Sources, Controlled Environments, and Sustainable Management

Alternative sources of water are available for agriculture, including treated wastewater, stormwater runoff, and water from energy production (produced water). It remains to be determined if contaminants in these waters (e.g., organic chemicals, heavy metals, salts, and radioactive elements) pose a significant risk to agriculture production, especially if they are deployed in the long term. In general, there is poor understanding of how to manage water systems for multiple purposes, which may ensure the availability of water for agriculture at the time it is needed and improve environmental quality and social welfare (Schoengold and Zilberman, 2007). The system-level consequences of using alternative sources of water for agriculture rather than through the existing treatment and disposal options have not been determined on a broad scale in the United States.

Water savings in CEA are proven; however, there are gaps in understanding of how these approaches affect overall system sustainability. The market potential and extent of cropping systems that can be utilized in CEA need to be determined. Currently, CEA and recirculating aquaculture systems are energy intensive so better energy management strategies are needed. The cost of the various CEA systems must also decrease to become competitive with traditional agriculture and imports. Also, the primary socioeconomic drivers that would promote acceptance of CEA need to be determined. The optimal approaches for lighting and cooling systems, temperature control, feed, disease management, and resilient systems (for aquaculture) and the optimal plant and fish traits for controlled environments still need to be determined to make these systems scalable, economically viable, and sustainable. There are still gaps related to coupling CEA systems to the other elements of the built environment to close the loop on nutrient and water use.

Alternative sources of water and controlled environments can improve water-use efficiency and water productivity. However, several research questions must be addressed to realize these opportunities:

1. What alternative water sources and system-level management practices are available for agriculture?
2. How can controlled environments be sustainably designed to lower water use and increase water productivity?

5. BARRIERS TO SUCCESS

5.1 Water Policies and Pricing

Adoption of water-saving opportunities and water management practices will require new policies and realistic water pricing for agriculture as well as the political will to enforce policies and prices across states and international boundaries. This includes revising "water rights" to encourage climate-appropriate crop choices and to optimize overall system performance with respect to water use. It may also require significant upgrades to water infrastructure to meet future water demands for agriculture. Water storage practices, interbasin transfers, and water distribution systems are currently inadequate to meet future water demands for agriculture. The lack of adequate water storage increases the vulnerability of the agricultural system to droughts, especially in arid regions.

5.2 Competing Demands for Water

Agriculture is the largest water consumer in the United States, but energy production also requires access to vast amounts of water. Given that agriculture is a high-volume, low-economic-value user of water, competing demands for water with higher economic value (such as energy production) may ultimately decrease the amount of water available for agriculture. This could prevent achieving the goal of using only renewable water for agriculture.

5.3 Limited Resources for Interdisciplinary Teams

Improving water-use efficiency in agriculture will require the simultaneous development and integration of multiple technologies and technological approaches from different research communities (see Box 6-1). Successful integration of these technologies will require adequate research funding for highly interdisciplinary teams focused on the opportunities described in this chapter. Currently, there are limited opportunities for the necessary convergence of disciplines needed to tackle these complex problems.

5.4 Public Acceptance of New Technologies for Water-Use Efficiency

Many new opportunities to improve water productivity will require new policies as well as consumer acceptance to implement direct water reuse, multiple-use scenarios, use of impaired waters, and integrated agro-ecosystems and natural ecosystems. New technologies (e.g., genetic modifications of plants using CRISPR-Cas9 or nanotechnology for soil and plant

BOX 6-1
Achieving Necessary Gains in Water-Use Efficiency

Changes in irrigation practices from flood irrigation to center-pivot sprinklers to drip irrigation have resulted in great improvements in water-use efficiency over the past several decades. Further gains will require overcoming systemic barriers to increasing water-use efficiency through the application of multiple technologies, improved data analytics, and better biophysical-based integrated systems models. Since water use in agriculture is sequential in nature, small gains in efficiency throughout each phase can produce considerable improvements in overall efficiency (Hsiao et al., 2007).

Increased efficiency could also be achieved through an ability to select from a number of genetic varieties of plants based on relevant weather and seasonal climate forecasts. Plants can be selected for varieties that are salt or drought tolerant in dry years when alternative impaired sources of water are used for irrigation. Hyperregional weather forecasts and sensors in sentinel plants along with ubiquitous soil moisture sensors enable application of water only when and where it is most needed. Accurate biophysical models, along with real-time water availability and weather monitoring at the basin or interbasin scale, are used to determine the appropriate source of water for that crop at that time. Taken together, these approaches can increase water-use efficiency from the plant to the basin scale.

sensors) are often viewed with skepticism and fear, which can be magnified when they are applied to food. Public and stakeholder engagement at all stages of technology development is needed to promote technology adoption. Research is needed to better understand the factors that would affect public acceptance of alternative water sources and new technologies.

5.5 Lack of a Systems Approach

A systems approach to water management is needed to mitigate the potential for unintended consequences to the natural ecosystem and built environments. Proposed interventions need to account for their impacts at large spatial scales (e.g., watershed, basin, or interbasin scales). The interactions between the elements of agroecosystems need to be better elucidated (see Chapter 8), and the development of appropriate models and systems-level analytics is needed in conjunction with technology advances for increasing water-use efficiency.

6. RECOMMENDATIONS FOR NEXT STEPS

Ensuring availability of water for agriculture is essential to meeting future demands. Emerging sensing technologies can provide real-time information about water availability and demand at high spatial and temporal resolution. Advances in data science and better integration of data and models will vastly improve the ability to predict water needs and optimize water-use efficiency. A better understanding of the soil properties affecting water availability and the physiological factors affecting plant water use—coupled with technologies that can modify soil, microbiome, and plant properties with respect to water use—will further enhance water-use efficiency. The use of controlled environments for some agriculture could meet the goal of making water use for agriculture sustainable. Realizing these opportunities will require a systems approach to water use and will require implementing a combination of these strategies simultaneously. Some high-priority opportunities for improving water-use efficiency include

- Increasing water-use efficiency by implementing multiple water-saving technologies across integrated systems;
- Lowering water use through applications of prescriptive analytics for water management;
- Lowering water demands by improving plant and soil properties to increase water-use efficiency; and
- Increasing water productivity by use of controlled environments and alternative water sources.

REFERENCES

Barbosa, G. L., F. D. Almeida Gadelha, N. Kublik, A. Proctor, L. Reichelm, E. Weissinger, G. M. Wohlleb, and R. U. Halden. 2015. Comparison of land, water, and energy requirements of lettuce grown using hydroponic vs. conventional agricultural methods. *International Journal of Environmental Research and Public Health* 12(6):6879-6891.

Block, P., and L. Goddard. 2012. Statistical and dynamical climate predictions to guide water resources in Ethiopia. *Journal of Water Resources Planning and Management* 138(3):287-298.

Bregnballe, J. 2015. *A Guide to Recirculation Aquaculture: An Introduction to the New Environmentally Friendly and Highly Productive Closed Fish Farming Systems.* Food and Agriculture Organization of the United Nations. Available at http://www.fao.org/3/a-i4626e.pdf (accessed June 24, 2018).

Brown, C., and U. Lall. 2006. Water and economic development: The role of variability and a framework for resilience. *Natural Resources Forum* 30(4):306-317.

CGLAMR (Center for Global Agricultural Monitoring Research, University of Maryland). 2018. *The Global Agriculture Monitoring (GLAM).* Available at http://glam.umd.edu/project/global-agriculture-monitoring-glam (accessed June 22, 2018).

DeLucia, E. H., N. Gomez-Casanovas, J. A. Greenberg, T. W. Hudiburg, I. B. Kantola, S. P. Long, A. D. Miller, D. R. Ort, and W. J. Parton. 2014. The theoretical limit to plant productivity. *Environmental Science & Technology* 48:9471-9477.

Giraldo, J. P., M. P. Landry, S. M. Faltermeier, T. P. McNicholas, N. M. Iverson, A. A. Boghossian, N. F. Reuel, A. J. Hilmer, F. Sen, J. A. Brew, and M. S. Strano. 2014. Plant nanobionics approach to augment photosynthesis and biochemical sensing. *Nature Materials* 13(4):400-408.

Glowacka, K., J. Kromdijk, K. Kucera, J. Xie, A. P. Cavanagh, L. Leonelli, A. D. B. Leakey, D. R. Ort, K. K. Niyogi, and S. P. Long. 2018. *Photosystem II Subunit S* overexpression increases the efficiency of water use in a field-grown crop. *Nature Communications* 868:1-9.

Harwood, E. 2017. Webinar presentation to the National Academies of Sciences, Engineering, and Medicine's Committee on Science Breakthroughs 2030: A Strategy for Food and Agricultural Research, November 10.

Hsiao, T. C., P. Steduto, and E. Fereres. 2007. A systematic and quantitative approach to improve water use efficiency in agriculture. *Irrigation Science* 25(3):209-231.

Karhu, K., T. Mattila, I. Bergström, and K. Regina. 2011. Biochar addition to agricultural soil increased CH_4 uptake and water holding capacity—results from a short-term pilot field study. *Agriculture, Ecosystems & Environment* 140(1-2):309-313.

Kisekka, I., and J. Aguilar. 2016. Deficit irrigation as a strategy to cope with declining groundwater supplies: Experiences from Kansas. Pp. 51-66 in *Emerging Issues in Groundwater Resources,* edited by A. Fares. Cham, Switzerland: Springer International.

Manzoni, S., J. P. Schimel, and A. Porporato. 2012. Responses of soil microbial communities to water stress: Results from a meta-analysis. *Ecology* 93(4):930-938.

Marasco, R., E. Rolli, B. Ettoumi, G. Vigani, F. Mapelli, S. Borin, A. F. Abou-Hadid, U. A. El-Behairy, C. Sorlini, A. Cherif, and G. Zocchi. 2012. A drought resistance-promoting microbiome is selected by root system under desert farming. *PLoS ONE* 7(10):e48479.

McGuinness, D. L., P. Pinheiro da Silva, E. W. Patton, and K. Chastain. 2014. Semantic eScience for Ecosystem Understanding and Monitoring: The Jefferson Project Case Study. American Geophysical Union, Fall Meeting 2014, Abstract ID. IN21B-3712.

Minasny, B., and A. B. McBratney. 2017. Limited effect of organic matter on soil available water capacity. *European Journal of Soil Science* 69(7):38-47.

Moshelion, M., O. Halperin, Ro. Wallach, R. Oren, and D. A. Way. 2015. Role of aquaporins in determining transpiration and photosynthesis in water-stressed plants: Crop water-use efficiency, growth and yield. *Plant Cell and Environment* 38(9):1785-1793.

NASEM (National Academies of Sciences, Engineering, and Medicine). 2016. *Next Generation Earth System Prediction: Strategies for Subseasonal to Seasonal Forecasts.* Washington, DC: The National Academies Press.

Olen, B., J. J. Wu, and Ch. Langpap. 2016. Irrigation decisions for major West Coast crops: Water scarcity and climatic determinants. *American Journal of Agricultural Economics* 98(1):254-275.

Pederson, G. T., S. T. Gray, C. A. Woodhouse, J. L. Betancourt, D. B. Fagre, J. S. Littell, E. Watson, B. H. Luckman, and L. J. Graumlich. 2011. The unusual nature of recent snowpack declines in the North American cordillera. *Science* 333(6040):332-335.

Schoengold, K., and D. Zilberman. 2007. The economics of water, irrigation, and development. *Handbook of Agricultural Economics* 3:2933-2977.

Stanhill, G. 1986. Water use efficiency. In *Advances in Agronomy* 39:53-85.

Steward, D. R., P. J. Bruss, X. Yang, S. A. Staggenborg, S. M. Welch, and M. D. Apley. 2013. Tapping unsustainable groundwater stores for agricultural production in the High Plains Aquifer of Kansas, projections to 2110. *Proceedings of the National Academy of Sciences of the United States of America* 110(37):E3477-E3486.

Stuart, D., B. Basso, S. Marquart-Pyatt, A. Reimer, G. P. Robertson, and J. Zhao. 2015. The need for a coupled human and natural systems understanding of agricultural nitrogen loss. *BioScience* 65(6):571-578.

Tilman, D., C. Balzer, J. Hill, and B. L. Befort. 2011. Global food demand and the sustainable intensification of agriculture. *Proceedings of the National Academy of Sciences of the United States of America* 108(50):20260-20264.

Turner, N. C. 2004. Agronomic options for improving rainfall-use efficiency of crops in dry-land farming systems. *Journal of Experimental Botany* 55(407):2413-2425.

Uddameri, V., and D. Reible. 2018. Food-energy-water nexus to mitigate sustainability challenges in a groundwater reliant agriculturally dominant environment (GRADE). *Environmental Progress & Sustainable Energy* 37(1):21-36.

USDA-ERS (U.S. Department of Agriculture's Economic Research Service). 2018. *Irrigation & Water Use.* Available at https://www.ers.usda.gov/topics/farm-practices-management/irrigation-water-use/#definitions (accessed February 7, 2017).

USGS (U.S. Geological Survey). 2016. *Irrigation Water Use.* Available at https://water.usgs.gov/watuse/wuir.html (accessed February 7, 2018).

Vengosh, A., R. B. Jackson, N. Warner, T. H. Darrah, and A. Kondash. 2014. A critical review of the risks to water resources from unconventional shale gas development and hydraulic fracturing in the United States. *Environmental Science & Technology* 48(15):8334-8348.

7

Data Science

1. INTRODUCTION

The growing availability of data presents an opportunity to improve the resilience and efficiency of food and agricultural production on a scale unimaginable even one decade ago. The convergence of cloud computing, mobile, Internet of Things (IoT), and analytic technologies has resulted in an explosion of data that has been transformational in all sectors of the economy (Cao, 2016). Approximately 90 percent of the data ever created were generated in the past 2 years (Marr, 2018). By 2025, the amount of data created and copied annually is expected to grow to 1 trillion gigabytes (Seagate, 2017). Understanding the connections between the biophysical and socioeconomic elements of agricultural systems through research depends on generating and analyzing massive amounts of data. Data will be at the center of the next revolution in food and agriculture (UN, 2014).

The relationship between data, algorithms, and computing is symbiotic. In practice, (raw) data are processed by computers through mathematical models and algorithms into useable information. The value of data comes when they are analyzed to provide information that is used to make decisions based on insights and understandings derived from data. Advances in data storage, communications, and processing have led to new research methods and tools that were simply not possible just one decade ago (NIH, 2018). Similarly, breakthroughs in agriculture will benefit from and compel state-of-the-art advances in the areas of data, algorithms, and computing. Machine learning and deep learning are two such examples, which will be further discussed in Section 3.1.

Data science is the emerging field that sits at the nexus of data, algorithms, and computing. It is an interdisciplinary field of inquiry in which quantitative and analytical approaches, processes, and systems are developed and used to extract knowledge and insights from increasingly large and complex sets of data (NIH, 2018). Data analysis has fostered knowledge creation for hundreds of years; achieving the science breakthroughs identified in this report will be spurred by data science.

In June 2018, the United States announced a new record in the global competition to build the world's fastest supercomputer. The petascale machine called "Summit" can perform 200 petaflops (10^{15} calculations) per second and is more than twice the speed of the incumbent at China's National Supercomputing Center (Top 500, 2018). The next horizon is exascale computing; a machine capable of executing 1 billion billion ($10^9 \times 10^9 = 10^{18}$) calculations per second. Exascale computing will enable more complete and accurate representations in Earth systems modeling needed to improve our understanding of local to global scales for the food system and enable faster simulations that can, for example, more realistically mirror the speeds of biological processes (McDermid et al., 2017).

Figure 7-1 shows a data analytics maturity curve that progresses from descriptive to predictive to prescriptive. In the early stages, information provided in the descriptive stage lends hindsight on what happened. The predictive stage is the next stage that provides insight into what will happen. The last stage of analytics maturity is prescriptive and enables foresight in what we can make happen. Much progress has been made in food and agricultural research in establishing descriptive analytics. Scientific breakthroughs in food and agricultural disciplines in the next 10-15 years will increasingly address the predictive and prescriptive levels of understanding.

Our understanding of the fundamental scientific underpinnings of the biological, chemical, physical, and socioeconomic elements of the food and agricultural system can benefit from better data access, data integration, and data analytics. The previous chapters have indicated specific instances where access to more data and data science can improve the resilience, efficiency, and sustainability of agriculture. This chapter describes the challenges of collecting, integrating, and analyzing a broad range of data types, and ensuring the quality of those data in near real time to afford the research breakthroughs needed. The chapter then proposes research needs and potential breakthroughs specific to data science for the food and agricultural enterprise.

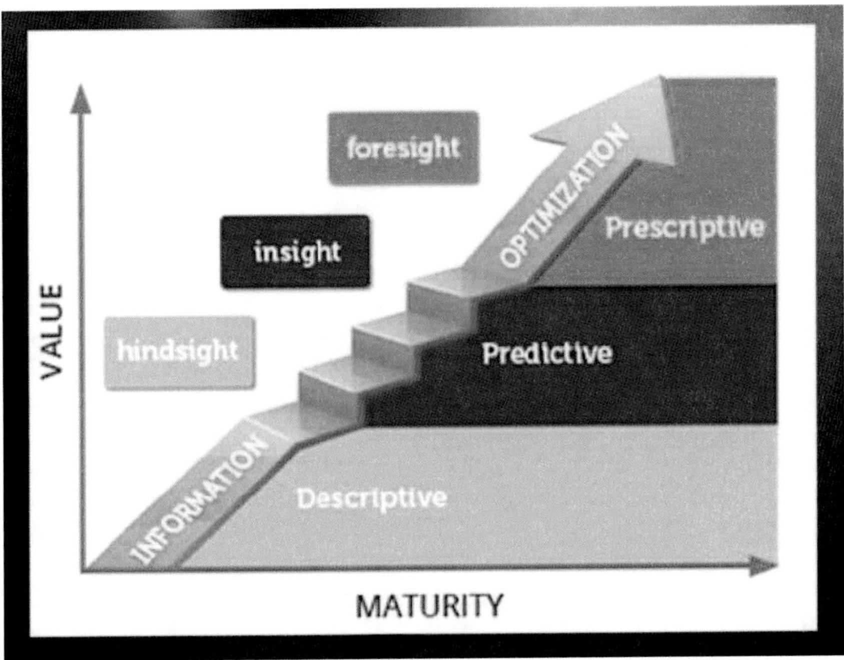

FIGURE 7-1 Analytics maturity curve.
SOURCES: Centurion, 2015, citing INFORMS Analytics Maturity Model User Guide (www.informs.org).

2. CHALLENGES

2.1 Challenge 1: Data Heterogeneity and Dimensionality

The challenges and opportunities in agricultural and food data can be illustrated by the acronym GEMS, which represents the fact that agriculture yields are modeled as a function of genetic (G), environment (E), management (M), and socioeconomic (S) factors. To understand yield, there needs to be an understanding of the genetics of the plant or animal being cultivated, the environmental factors affecting growth, the management practices of the farm, and the socioeconomic factors, as well as an understanding of the complex interactions among these factors, represented as G × E × M × S (GEMS). Numerous environmental factors may be of interest (e.g., location, soil type, elevation, inclination, rainfall, precipitation, humidity, temperature, and dew point) and each may have positive or negative impacts on genotypes. GEMS data are in a wide variety of formats, in different spatial and temporal scales, with different degrees of accuracy and precision that are challenging to harmonize for analysis and discovery (Lu et al., 2016). Data curation and harmonization efforts currently constitute 80 percent of the effort related to data use (Crowdflower, 2016).

The curation and analysis of GEMS data lead to the need for data standards and tools to manipulate and analyze the massive geospatial temporal datasets, and pushes the frontier of data, analytics, and computing systems (see Box 7-1). The complexities of data, such as the GEMS data, can be explored in terms of the four basic data attributes: volume, variety, velocity, and veracity (see Box 7-2).

Example: Improvements in Sensor Technology and Data Velocity Enabling Real-Time Continuous Monitoring in Agriculture. Novel sensors can provide accurate measures of soil, plant, animal, and food characteristics at a high frequency and data density and at any given time and location. The next generation of sensors will revolutionize the ability to deploy prescriptive solutions in near real time. For example, Figure 7-2 shows how georeferenced in situ soil sensors could be widely distributed across a farm, measure soil and plant nutrients, and display results in a colored geospatial map that allows farmers to determine site-specific actions (e.g., variable application of water and nutrients). Similarly, microelectromechanical technology is already being used to develop miniature sensors that

BOX 7-1
G.E.M.S.™

G.E.M.S.™ (Genetic, Environmental, Management, and Socioeconomic data) is a proprietary platform developed by the University of Minnesota that merges domain expertise in the food and agricultural sciences with high-performance computing and bioinformatics expertise to drive the development of next-generation agricultural and food informatics (agri-food informatics), data discovery, and analysis. G.E.M.S.™ aims to allow researchers to solve problems at multiple functional, temporal, and spatial scales related to crop sustainability and food production through provision of interoperable genetic, environmental, management, and socioeconomic data related to agriculture.

There are massive datasets involved with each of these components, and a data lake (or a data repository) is required to hold data until they are needed. Currently there are not standards or application programming interfaces to allow those datasets to integrate and merge with one another to be useful.

SOURCE: Allan et al., 2017.

BOX 7-2
Attributes of Data

Volume. Data volume is the size of the data and is typically measured in bytes. Genomic experiments generate terabytes (TB) of data. For example, the genome of a single corn plant consists of 2.5 billion DNA bases (USDA-NIFA, 2017a). By 2025, genomics is expected to be on par with or exceed the data needs of the Big Data domains of astronomy, YouTube, and Twitter (Stephens et al., 2015).

Variety. Data variety refers to the number of different types of data. Precision agriculture requires many different layers of data to develop a site-specific understanding. For example, variables could include soil type, elevation, slope, temperature, and precipitation. Satellite data can provide critical real-time information, but the heterogeneity (or variety) of data makes it difficult to combine the data in a harmonized way needed for analytical methods to examine specific factors for research or for decisions. Amassing spatial-temporal data in a harmonized way would be transformational for real-time decision support (e.g., disease prediction via disease vectors using spatial data).

Velocity. Data velocity indicates the speed of data processing. With rapid improvements in the speed of data velocity, real-time continuous monitoring of agriculture is becoming possible.

Veracity. Data veracity refers to bias, noise, abnormality, uncertainty, and imprecision. This often-overlooked dimension of data is important in developing appropriate mathematical models and analytics to obtain useful insights. The food and agricultural system is an intricate system of systems with multiple dimensions of uncertainty; for instance, at a macro level, agriculture has both demand and supply uncertainty (IOM and NRC, 2015).

can be embedded into plant stems for measuring a plant's water potential (the hydration relative to its growth and production yield) (Pagay et al., 2014); this could replace traditional labor-intensive, destructive methods that provide only point-in-time measurements. Continuous measurements such as these will improve the calibration and performance of water- and nutrient-use models, and may lead to a new understanding of how plants use water and nutrients. Integrating these farm-scale sensor technologies with seasonal or hyper-local weather forecasts, or with measures of water availability in connected ecosystems via the IoT, creates opportunities to manage farm ecosystems integrated with natural ecosystems (e.g., ecohydrology). For vineyards, such an understanding could lead to more consistent production of high-quality wine grapes with certain flavor and

FIGURE 7-2 Schematic representation of real-time continuous monitoring of soil, crops, water, and livestock via microelectromechanical and biodegradable sensors. The sensors measure a range of important attributes to generate data. Data can then be analyzed and used for modeling and for predictive and prescriptive agriculture.

aroma profiles (Campbell, 2016). Likewise, in planta crop health sensors may quantify biochemical changes in plants caused by an insect pest or a pathogen, alerting and enabling the producer to plan and deploy immediate site-specific control strategies before the infestation occurs and damage is visible and widespread across the field. Increased data velocity will enable dynamic control of agricultural equipment in motion in real time—such as precision planters, sprayers, and irrigation—and enable real-time continuous monitoring of individual livestock in a herd using wearable (and other) sensors for precision livestock applications, using advanced technologies such as microfluidics, sound analyzers, image-detection techniques, sweat and salivary sensing, serodiagnosis (diagnosis based on blood sera), and others. Creating efficient online monitoring systems in real time requires the ability to integrate all the available sensor data and run analysis in real time (Neethirajan, 2017).

2.2 Challenge 2: Data Standards and Interoperability

Modern scientific research is becoming a more data-driven, interactive process between multiple stakeholders across the world (EU SCAR, 2015). The term "e-science" has been coined to describe the potential of data-driven and computationally intensive processes to enhance manual laboratory work and fieldwork and reduce dependence on paper-based recording (EU SCAR, 2015). E-science has the potential to enable worldwide collaboration in flexible teams using advanced tools, services, and data repositories. These can include distributed networks or grid computing, high-performance computing, visualizations, simulations, workflows, and provenance documentation (EU SCAR, 2015).

Providing access to data can accelerate and democratize the scientific process. To do so would require infrastructure improvements for supporting the reuse of scientific data in the food and agricultural realm and enhancements for machines to automatically find and use data. While there are some notable cyberinfrastructure initiatives and data-sharing efforts under way (see Box 7-3), these efforts are only able to address some parts of the vast food system. As a set of universally agreed-upon guidelines referring to interoperability among systems or applications, standards support reuse. One of the major hurdles in data standardization and interoperability is the lack of sufficient knowledge representations (e.g., ontologies, semantic nets, and rules) (Jonquet et al., 2018). Knowledge representations provide a representation that a computer system can utilize to solve complex tasks.

The FAIR data principles are a set of guiding principles that could facilitate data standardization and interoperability for scientific data management and stewardships. The principles are organized around four concepts: findable, accessible, interoperable, and reuseable (Wilkinson et al., 2016). The FAIR principles are complementary to open-data philosophy—data can be open without being FAIR, and vice versa. Open data is the idea that some data should be freely available to everyone to use and republish as they wish, without restrictions from copyright, patents, or other mechanisms of control (Auer et al., 2007). Government-funded awards often require researchers to open their data, but there is little guidance on best practices or dissemination for open data in the United States.

The existence of standards is an often-cited success factor for reuse of data and software (Pasquetto et al., 2017). Standards can be administered in a variety of ways: created de facto by private organizations[1] or managed by nonprofit organizations.[2] The nonprofit consortium of businesses serv-

[1]The computer file format PDF was created by Adobe in 1993 and was a de facto standard for many years before eventually becoming a formal ISO standard in 2005.

[2]The nonprofit organization GS1 is best known for the barcode (universal product code) that is scanned by retailers more than 6 billion times every day.

BOX 7-3
Data Sharing and Cyberinfrastructure Initiatives Related to Food and Agriculture

Below are examples of some food and agricultural data sharing and cyberinfrastructure initiatives that are currently under way.

- INFEWS (Innovations at the Nexus of Food, Energy and Water Systems)[a] is a National Science Foundation (NSF) program that seeks to support research that conceptualizes Food-Energy-Water systems broadly and inclusively, including social and behavioral processes (such as decision making and governance), physical processes (such as built infrastructure and new technologies for more efficient resource utilization), natural processes (such as biogeochemical and hydrologic cycles), biological processes (such as agroecosystem structure and productivity), and cybercomponents (such as sensing, networking, computation, and visualization for decision making and assessment). Through funding from NSF and the U.S. Department of Agriculture's (USDA's) National Institute of Food and Agriculture, the initiative aims to identify and fund the most meritorious and highest-impact projects across the food–energy–water nexus. The initiative also promotes international cooperation among scientists and engineers from a range of disciplines and organizations.
- CyVerse[b] is an NSF-funded project that aims "to design, deploy and expand a national cyberinfrastructure for life sciences research and to train scientists in its use." The project provides life scientists with the computational infrastructure to manage large datasets and complex analyses through data storage, bioinformatics tools, image analyses, cloud services, APIs, and other tools. CyVerse is a dynamic virtual organization led by the University of Arizona alongside the Texas Advanced Computing Center, Cold Spring Harbor Laboratory, and the University of North Carolina at Wilmington.
- USDA VegScape[c] is a National Agricultural Statistical Service (NASS) web service–based U.S. crop condition monitoring system. It aims to improve

ing the agriculture sector, AgGateway, has identified the need for standards as essential to promoting and enabling the industry's transition to highly data-driven agricultural practices and is working to develop their own (AgGateway, 2017; Smith, 2017).

2.3 Challenge 3: Data Privacy

Privacy is a confounding issue for food and agricultural data. In some domains, such as health care, data can be anonymized by removing or encrypting personally identifiable information. The best available privacy protection technology, differential privacy, is inadequate for agricultural objectivity, robustness, quantification, and defensibility of nationwide crop condition monitoring by delivering interactive vegetation indexes that enable users to explore, visualize, query, and disseminate current vegetative cover maps and data. New satellite-based data are loaded on a weekly basis during the growing season. VegScape is built on the CropScape[d] (a NASS geospatial cropland data service) framework and uses NASA's MODIS satellite (Yang et al., 2013).

- GEOGLAM (International Group on Earth Observations Global Agricultural Monitoring Initiative)[e] is an initiative developed by the Group on Earth Observations (a partnerships of governments and international organizations) which coordinates satellite monitoring observation systems in different regions of the world in order to enhance crop production projections and weather forecasting data. The initiative provides a framework using Earth observations, including satellite and ground-based observations, designed to build on existing agricultural monitoring programs and initiatives at national, regional, and global levels.
- AgMiP (the Agricultural Model Intercomparison and Improvement Project)[f] is an international project that utilizes intercomparisons of various methodologies to "improve crop and economic models and ensemble projections and to produce enhanced assessments by the crop and economic modeling communities researching climate change agricultural impacts and adaptation." AgMiP is using a two-track science approach of model intercomparison and improvement and climate change multimodel assessment. These approaches are facilitated by a series of regional workshops held in each AgMiP region over a 3-year period as well as by global studies and workshops that focus on particular crops and on global analyses.

---

[a]See https://www.nsf.gov/funding/pgm_summ.jsp?pims_id=505241.
[b]See http://www.cyverse.org/about.
[c]See https://nassgeodata.gmu.edu/VegScape.
[d]See https://nassgeodata.gmu.edu/CropScape.
[e]See https://cropmonitor.org.
[f]See http://www.agmip.org.

data because of its spatiotemporal nature (Shekhar et al., 2017). Anonymizing geospatial data without distorting its granularity is a known problem which does not yet have an adequate solution. Aggregating data is a frequently used approach to anonymize farm- or field-level data. For instance, data from individual fields and farms are aggregated into county-level metrics that are made public. Aggregation limits the granularity of analysis, making intrafield or site-specific analysis impossible. For example, spatial data that are aggregated for anonymization purposes typically lack the high spatial resolution needed by biophysics-based models to accurately capture or predict the system responses.

3. SCIENTIFIC OPPORTUNITIES

There are emerging technologies related to data that can transform the food and agricultural system. In particular, there are four promising future areas: artificial intelligence, blockchain, IoT, and quantum computing.

3.1 Opportunity 1: Artificial Intelligence

Artificial intelligence (AI) refers to intelligence displayed by machines, as opposed to natural intelligence demonstrated by humans and other animals (Bradford, 2017). The term AI is commonly used when a machine mimics cognitive functions that are typically associated with humans, such as planning, learning, reasoning, problem solving, knowledge representation, perception, motion, manipulation, social intelligence, and creativity. AI combines automation, robotics, and computer vision. Advances in statistics, faster computers, and access to large amounts of data have enabled advances in AI, particularly in the field of machine learning where significant progress has been made in the areas of image and pattern recognition, natural language understanding, and robotics.

The scale and complexity of the food and agricultural system make it well suited for AI, with large amounts of data available to train algorithms. Large amounts of data from ever-more-sophisticated experiments and models are being generated, and food and agricultural researchers are challenged with deciding what data to collect and how to best process the data generated. A 2016 analysis estimated that a new crop protection product requires the analysis of more than 160,000 compounds—the equivalent to more than 11 years of research and development and overall costs that exceed $280 million per commercial product (Brayne et al., 2018). AI will enable new discoveries, not just in analyzing existing data but in augmenting human creativity around planning new experiments and accelerating discovery and improving efficiency.

There are promising applications and indicators for AI in agriculture. In selective breeding programs, the development of high-throughput automated phenotyping capabilities can speed the process of breeding to develop resilient yet high-yielding, high-quality crops. AI and machine-learning research on the massive amount of environmental, growth, and genetic data can help scientists, for example, elucidate the connections between system components, predict the yield of different varieties based on early-season plant attributes, and associate specific desirable traits with genetic markers (Baxter et al., 2007). Robots can be designed to harvest crops at a higher volume and faster pace than human laborers, to monitor crop and soil health using computer vision and deep-learning algorithms that analyze data captured by drones, and to more accurately predict crop

BOX 7-4
Robotic Milking

One agricultural application that simultaneously utilizes and collects large datasets is robotic milking systems (RMSs). Using detection sensors, the robot attaches the cups to the cow teats from underneath the udder. These sensors are able to detect how many teats are available and their positioning to ensure that the cups are attached correctly. Every cow wears an electronic tag making it possible for the RMS to identify each cow and collect production and other data on her. Cows are only allowed access to be milked if a certain period of time has elapsed since their last milking. A single stall can milk 55-65 cows per day, and it is estimated that more than 35,000 RMS units are operational on dairy farms around the world. RMS results in a 5-10 percent increase in milk yield as compared to milking two times per day, mainly due to increased milking frequency (de Koning, 2010).

A milk analyzer can be included in the RMS to collect data on milk components such as fat, protein, pH, and lactose. These data can provide real-time alerts to notify the farmer of feeding or health problems or signal the need for a management intervention. Furthermore, machine learning would provide an opportunity to tune parameters for assessing milk quality. The profitability of RMSs depends on factors such as the amount of milk produced per cow and per robot, labor savings, and length of useful life (Salfer et al., 2017). Although RMSs typically have not been profitable in U.S. dairy production systems to date, advances in robotic design, availability of labor, and higher labor costs in the future may alter this situation, as has been the case in western and northern Europe, which have the highest rates of RMS adoption (Brou ek and Tonge , 2015).

yields using machine-learning models to better understand environmental impacts (Sennar, 2017). Box 7-4 on robotic milking provides an example of integrating AI with sensor technologies in animal agriculture.

3.2 Opportunity 2: Blockchain

Blockchain is a recent technology advancement with potential for addressing the challenge of creating a more transparent, authentic, and trustworthy digital record of the journey that food and other physical products take across the supply chain (Lougee, 2017). Blockchain technology arose out of the efforts to create the cybercurrency Bitcoin, but its potential goes well beyond its original purpose. As discussed in Chapter 4, at its core, the blockchain is a shared, immutable ledger for recording the history of transactions (IEEE, 2018a). It uses a variety of technologies, including public–private key cryptography, distributed databases, decentralized processing, hash functions, and consensus algorithms (Ge et al., 2017).

This could be valuable in creating transparency in a global food system with multiple disparate actors across the multitiered supply chain.

More information will be required to address research challenges on food and agricultural supply chains and a platform for studying how complex systems interact. Blockchain presents a potential new source of data and new platform to be leveraged to deliver advances toward transparency to the food system and transformations in food safety, fraud reduction, market access, waste reduction, and productivity gains (Ahmed and ten Broek, 2017; Ge et al., 2017; Kim and Laskowski, 2017).

A systematic mapping study designed to understand the current research topics, challenges, and future directions of blockchain technology from a technical perspective examined 41 primary papers (Yli-Huumo, 2016). It showed that most of the current research on blockchain technologies has been focused on security and privacy issues. Issues in scalability, such as performance and latency, need to be addressed in order to realize pervasive use of blockchain technology in the food system.

3.3 Opportunity 3: IoT

IoT is the network of physical devices embedded with electronics, software, sensors, actuators, and connectivity which enables these "things" to connect and exchange data creating opportunities for more direct integration of the physical world into computer-based systems, resulting in efficiency improvements, economic benefits, and reduced human exertions (Wasik, 2013; Morgan, 2014). Among its advantages, IoT addresses the high cost of manual data collection, which impedes the adoption of beneficial data-driven technologies. IoT technologies enable seamless data collection from various sensors, cameras, and drones and are the foundation for intelligent systems, as illustrated in the example above and throughout this report. Farm applications of precision agriculture and precision livestock farming are prime examples of IoT; however, the influence of IoT spans the entire food system (Wolfert et al., 2017).

IoT is still in its infancy in the food and agriculture sector (Verdow et al., 2016). Realizing future subsystems within the food system where IoT devices are capturing and sending data, but in a time-sensitive and synchronized way, "could be stalled by our lack of effective methods to marry computers and networks with timing systems" (NIST, 2015).

Challenging requirements for food and agriculture applications need to be addressed in both the technical and nontechnical arenas. For example, on-site applications in remote outdoor farming locations often lack Internet connectivity needed for communications. Easy access to power and durability and calibration are also essential. The FarmBeats system design explicitly accounts for weather-related power and Internet outages, which has enabled 6-month-long deployments in two U.S. farms (Vasisht et al., 2017).

3.4 Opportunity 4: Quantum Computing

Quantum mechanics may play a vital role in biology, as quantum phenomena occur in biological systems. In 2007, it was discovered that photosynthesis operates though quantum coherence, giving rise to the new field of quantum biology (Engel et al., 2007; Ball, 2011). Other examples of quantum phenomena relevant for food and agriculture include olfaction (Turin, 2002), DNA mutation (Lowdin, 1965), and Brownian motors in many cellular processes (Krug et al., 2006). The emergence of quantum computing could be used to explore natural phenomena in the physical world—phenomena that traditional computing paradigms are ill suited to represent at scale.

Quantum mechanics is a subdiscipline of physics that explains the physical world and attempts to explore how it functions (Feynman, 2002). Nature follows the laws of quantum mechanics, in which particles behave in "strange" ways. For example, caffeine molecules in a cup of coffee are difficult to model, and the detailed structure and properties are difficult to understand because the particles can take on more than one state and can even interact with other particles that are far away. The computers currently in use have not been able to take into account such complex properties because classical computers code information in bits that represent 0 or 1 values. On the other hand, quantum computers operate on the basis of qubits and incorporate two key principles of quantum physics: superposition and entanglement. Superposition allows a qubit to simultaneously take on the value of both 0 and 1, and entanglement allows qubits in a superposition to be correlated with one another (the state of one can be dependent on the other). Classical computers use a binary format as on-off switches, while quantum computers use qubits that can act as sophisticated on-off switches and can be used to solve more complex problems in quantum scales (including nature) (IEEE, 2018b).

Quantum computing technology is in its infancy. Many fundamental challenges exist, including developing long-living qubits for computation, scaling the number of qubit processors, and building effective quantum error correction. However, major efforts are under way across the globe in academia and industry and promising early results are unfolding (Knight, 2017; Alibaba, 2018; Google, 2018; IBM Q, 2018; Intel, 2018; Microsoft, 2018). In 2016, the first publicly available quantum computing service on the cloud was made available. Quantum computers may lead to revolutionary breakthroughs in discovery of new materials for agriculture, agrochemical discovery, and artificial intelligence algorithms that could impact food and agriculture. Molecular modeling has been identified as an area for exploration with quantum computing (Emerging Technology from the arXiv, 2017). Another area in which quantum computing could help with is analyzing genomics data for plants, animals, or the soil microbiome. Quantum computing could make it possible to perform real-time analytics on various types of data (e.g., weather data at high spatial resolution) (Accenture Labs, 2017; Popkin, 2017). This unique juncture in time provides an opportunity for food and agricultural research to help shape the next era of computing.

4. BARRIERS TO SUCCESS

Data science is a rapidly evolving field and data science skills for working with Big Data are in high demand in all sectors, including food and agricultural research and agricultural economics (Woodward, 2016). Data science tools need to become more useable by non–computer science domain experts. Traditional agricultural and food science programs are evolving to better integrate the variety of disciplines needed; for instance, the digital agriculture initiative at Cornell University is aimed at generating innovating research at the intersection of agriculture, computing, and engineering (Cornell University, 2018). New partnerships, prizes, and conferences are needed to spark convergence between those who are new to food and agriculture and current food and agricultural researchers. Initiatives such as the Syngenta Crop Challenge in Analytics awarded by the Analytics Society of INFORMS show promise in fostering much-needed cross-industry collaboration (INFORMS, 2018). Reward systems for generating and sharing data in research environments may need to be considered, for example, creating incentive structures or changing the approach to tenure and promotion evaluations (USDA-NIFA, 2017b). It will be important to train a workforce that can manage, analyze, and manipulate large datasets and enable the workforce's convergence with food and agricultural scientists. Furthermore, it may be necessary to develop a culture that supports and rewards sharing of data (including combinations of private and public data) by sets of communities of researchers, standardizes protocols, harmonizes experimental designs, and addresses ownership, privacy, and security concerns unique to the food and agricultural enterprise. Funding sources, public or private, need to be available for developing the standards and protocols that need to be harmonized for agricultural datasets to make them interoperable.

The collection of data from numerous stakeholders is required to create the pool of Big Data needed to well represent the multidimensional food system. To incentivize data sharing by individual stakeholders or researchers, reward systems may require new technologies that can identify and appropriately attribute value derived from the pool of Big Data to the individual contributions. These value attribution technologies may be conceptually similar in spirit to marketing attribution, the process of identifying a set of user actions (events, touchpoint) that contribute in some manner to a desired outcome, and then assigning a value to each of these (Priest, 2018), but will need to account for the unique complexities in the food and agricultural domain.

Ownership of food and agricultural data can be a confounding issue, especially when it comes to IoT data. Ownership is a legal concept related to property, and U.S. law recognizes various categories of property (for example, real property, personal property, and intellectual property). However, data generated in the food and agricultural sectors may not belong in any of those categories of property. For example, farming data is a compilation of data generated from field operations (real property) using sensors on equipment (personal property) and represents intellectual property on maximizing yield. Questions emerge around legal ownership of data (which confers control) and how the data can be used. There are currently few legislative or judicial rulings for guidance. The American Farm Bureau Federation developed the *Privacy and Security Principles for Farm Data* (AFBF, 2014), which lays out 13 data principles, but these are only guidelines. Users of data will need to consider the implications of data ownership, including protecting ownership of data, abiding by legal usage, and ensuring appropriate sharing with others. Ambiguity and uncertainty around ownership can cause inefficiencies and limit sharing and value creation.

5. RECOMMENDATIONS

Advances in data sciences can transform how data can be better collected, analyzed, and used for food and agricultural research. While there are many opportunities, the following actions merit high priority:

- Accelerate innovation by building a robust digital infrastructure that houses and provides FAIR (findable, accessible, interoperable, and reuseable) and open access to agri-food datasets.
- Develop a strategy for data science in food and agricultural research, and nurture the emerging area of agri-food informatics by adopting and influencing new developments in data science and information technology in food and agricultural research.
- Address privacy concerns and incentivize sharing of public, private, and syndicated data across the food and agricultural enterprise by investing anonymization, value attribution, and related technologies.

REFERENCES

Accenture Labs. 2017. Innovating with Quantum Computing: Enterprise Experimentation Provides View into Future of Computing. Available at https://www.accenture.com/t00010101T0000000w/br-pt/acnmedia/PDF-45/Accenture-Innovating-Quantum-Computing-Novo.pdf (accessed April 26, 2018).

AFBF (American Farm Bureau Federation). 2014. Privacy and Security Principles for Farm Data. Available at https://www.fb.org/issues/technology/data-privacy/privacy-and-security-principles-for-farm-data (accessed April 26, 2018).

AgGateway. 2017. AgGateway Releases Annual Report and 5-Year Plan: Enable Companies to Increase Efficiency, Agility and Profitability. Available at http://www.aggateway.org/Newsroom/2017PressReleases/AgGatewayReleasesAnnualReportand5-YearPlan.aspx (accessed April 26, 2018).

Ahmed, S., and N. ten Broek. 2017. Food supply: Blockchain could boost food security. *Nature* 550(7674):43.

Alibaba. 2018. Alibaba Cloud and CAS Launch One of the World's Most Powerful Public Public Quantum Computing Services. Available at https://www.alibabacloud.com/press-room/alibaba-cloud-and-cas-launch-one-of-the-worlds-most (accessed July 9, 2018).

Allan, G., J. Erdmann, A. Gustafson, A. Joglekar, M. Milligan, G. Onsongo, K. Pamulaparthy, P. Pardey, T. Prather, S. Senay, K. Silverstein, J. Wilgenbusch, Y. Zhang, and P. Zhou. 2017. G.E.M.S: An innovative agroinformatics data discovery and analysis platform. Available at https://rdmi.uchicago.edu/papers/08162017165531paperwilgenbusch081617.pdf (accessed July 9, 2018).

Auer, S. R., C. Bizer, G. Kobilarov, J. Lehmann, R. Cyganiak, and Z. Ives, Z. 2007. DBpedia: A Nucleus for a Web of Open Data. *The Semantic Web*. Lecture Notes in Computer Science. 4825. P. 722. doi: 10.1007/978-3-540-76298-0_52.

Ball, P. 2011. Physics of life: The dawn of quantum biology. *Nature* 474:272-274.

Baxter, I., M. Ouzzani, S. Orcun, B. Kennedy, S. S. Jandhyala, and D. E. Salt. 2007. Purdue Ionomics Information Management System. An integrated functional genomics platform. *Plant Physiology* 143(2):600-611.

Bradford, A. 2017. Empirical evidence: A Definition. LiveScience. Available at https://www.livescience.com/21456-empirical-evidence-a-definition.html (accessed July 9, 2018).

Brayne, S., S. McKellar, and K. Tzafestas. 2018. Artificial Intelligence in the Life Sciences & Patent Analytics: Market Developments and Intellectual Property Landscape. London: IP Pragmatics Ltd. Available at https://www.ip-pragmatics.com/media/1049/ip-pragmatics-artificial-intelligence-white-paper.pdf (accessed July 9, 2018).

Brouček, J., and P. Tongel'. 2015. Adaptability of dairy cows to robotic milking: A review. *Slovak Journal of Animal Science* 48(2):86-95.

Campbell, G. 2016. The Tensiometer: Micro-sized. Environmental Biophysics. Available at http://www.environmentalbiophysics.org/tensiometers-micro-sized (accessed July 9, 2018).

Cao, L. 2016. Data science and analytics: A new era. *International Journal of Data Science and Analytics* 1(1):1-2.

Centurion, C. 2015. Moving Along the Analytics Maturity Curve. River Logic Blog. Available at https://blog.riverlogic.com/moving-along-the-analytics-maturity-curve (accessed July 9, 2018).

Cornell University. 2018. Digital Agriculture: Cornell University Agricultural Experiment Station. Available at https://cuaes.cals.cornell.edu/digital-agriculture (accessed April 26, 2018).

Crowdflower. 2016. Data Science Report. Available at http://visit.crowdflower.com/rs/416-ZBE-142/images/CrowdFlower_DataScienceReport_2016.pdf (accessed July 9, 2018).

de Koning, K. 2010. Automatic milking—Common practice on dairy farms. Pp. V59-V63 in *Proceedings of the First North American Conference on Robotic Milking*. Elora, ON, Canada: Precision Dairy Operators.

Emerging Technology from the arXiv. 2017. Google Reveals Blueprint for Quantum Supremacy. MIT Technology Review. Available at https://www.technologyreview.com/s/609035/google-reveals-blueprint-for-quantum-supremacy (accessed July 9, 2018).

Engel, G. S., T. R. Calhoun, E. L. Read, T. K. Ahn, T. Mancal, Y. C. Cheng, R. E. Blakenship, and G. R. Fleming. 2007. Evidence for wavelike energy transfer through quantum coherence in photosynthetic systems. *Nature* 466:782-786.

EU SCAR (European Union Standing Committee on Agricultural Research). 2015. Agriculture Knowledge and Innovations Systems Towards the Future: A Foresight Paper. Strategic Working Group AKIS-3 Report. Available at https://ec.europa.eu/research/scar/pdf/akis-3_end_report.pdf (accessed March 15, 2018).

Feynman, R. 2002. Richard Feynman on quantum physics and computer simulation. *Los Alamos Science* No. 27. Los Alamos National Laboratory. Available at http://permalink.lanl.gov/object/tr?what=info:lanl-repo/lareport/LA-UR-02-4969-02 (accessed July 9, 2018).

Ge, L., C. Brewster, J. Spek, A. Smeenk, and J. Top. 2017. Blockchain for Agriculture and Food: Findings from the Pilot Study. Wageningen Economic Research. Available at https://www.wur.nl/upload_mm/b/3/d/df37c4ff-14f3-43b9-a34d-6ed8599d8aba_2017-112%20Ge_def.pdf (accessed July 9, 2018).

Google. 2018. A Preview of Bristlecone: Google's New Quantum Processor. Available at https://ai.googleblog.com/2018/03/a-preview-of-bristlecone-googles-new.html (accessed July 9, 2018).

IBM Q. 2018. What Is Quantum Computing? Available at https://www.research.ibm.com/ibm-q/learn/what-is-quantum-computing (accessed March 15, 2018).

IEEE (Institute of Electrical and Electronics Engineers). 2018a. Blockchain Overview. Available at https://blockchain.ieee.org/about (accessed March 15, 2018).

IEEE. 2018b. Quantum Computers Strive to Break Out of the Lab. Available at https://spectrum.ieee.org/computing/hardware/quantum-computers-strive-to-break-out-of-the-lab (accessed March 15, 2018).

INFORMS (Institute for Operations Research and the Management Sciences). 2018. Syngenta Crop Challenge. Available at http://connect.informs.org/analytics/news-events/news-events-articles/news-events-news-events-articles-syngenta (accessed April 26, 2018).

Intel. 2018. Intel Starts Testing Smallest "Sping Qubit" Chip for Quantum Computing. Available at https://www.intc.com/investor-relations/investor-education-and-news/investor-news/press-release-details/2018/Intel-Starts-Testing-Smallest-Spin-Qubit-Chip-for-Quantum-Computing/default.aspx (accessed March 15, 2018).

IOM and NRC (Institute of Medicine and National Research Council). 2015. *A Framework for Assessing Effects of the Food System*. Washington, DC: The National Academies Press.

Jonquet, C., A. Toulet, E. Amund, S. Aubin, E. D. Yeumo, V. Emonet, J. Graybeal, M. Laporte, M. A. Musen, V. Pesce, and P. Larmande. 2018. AgroPortal: A vocabulary and ontology repository for agronomy. *Computers and Electronics in Agriculture* 144:126-143.

Kim, H. M., and M. Laskowski. 2017. Agriculture on the blockchain: sustainable solutions for food, farmers, and financing. Available at http://dx.doi.org/10.2139/ssrn.3028164.

Knight, W. 2017. IBM Raises the Bar with a 50-Qubit Quantum Computer. MIT Technology Review. Available at https://www.technologyreview.com/s/609451/ibm-raises-the-bar-with-a-50-qubit-quantum-computer (accessed March 15, 2018).

Krug, H., H. Brune, G. Schmid, U. Simon, V. Vogel, D. Wyrwa, H. Ernst, A. Grunwald, W. Grunwald, and H. Hofmann. 2006. *Nanotechnology: Assessment and Perspectives*. Springer-Verlag Berlin and Heidelberg GmbH & Co. K. Pp. 197-240.

Lougee, R. 2017. Are "Bytes" and "Blocks" the Secret Ingredients to Transforming Food Safety? SwissRe Institute. Available at http://institute.swissre.com/research/library/Food_Safety_Robin_Lougee.html (accessed March 15, 2018).

Lowdin, P. 1965. Quantum genetics and the aperiodic solid. Some aspects on the biological problems of heredity, mutations, aging and tumours in view of the quantum theory of the DNA molecule. *Advances in Quantum Chemistry.* Volume 2. Pp. 213-360. Academic Press.

Lu, S., L Shao, M. Freitag, L. Klein, J. Renwick, F. Marianno, C. Albrecht, and H. Hamann. 2016. IBM PAIRS Curated Big Data Service for Accelerated Geospatial Data Analytics and Discover, 2016 IEEE International Conference on Big Data. Available at https://ieeexplore.ieee.org/document/7840910 (accessed March 15, 2018).

Marr, B. 2018. How Much Data Do We Create Every Day? The Mind-Blowing Stats Everyone Should Read. Available at https://www.forbes.com/sites/bernardmarr/2018/05/21/how-much-data-do-we-create-every-day-the-mind-blowing-stats-everyone-should-read/#1d51008860ba (accessed July 9, 2018).

McDermid, S., L. Mearns, and A. Ruane. 2017. Representing agriculture in Earth systems models: Approaches and priorities for development. *Journal of Advances* 9(5):2230-2265.

Microsoft. 2018. Microsoft Quantum: Research. Available at https://www.microsoft.com/en-us/research/lab/quantum (accessed March 15, 2018).

Morgan, J. 2014. A Simple Explanation of "The Internet of Things." Available at https://www.forbes.com/sites/jacobmorgan/2014/05/13/simple-explanation-internet-things-that-anyone-can-understand/#958725c1d091 (accessed July 9, 2018).

Neethirajan, S. 2017. Recent advances in wearable sensors for animal health management. *Sensing and Bio-Sensing Research* 12:15-29.

NIH (National Institutes of Health). 2018. *NIH Strategic Plan for Data Science.* Available at https://commonfund.nih.gov/sites/default/files/NIH_Strategic_Plan_for_Data_Science_Final_508.pdf (accessed March 15, 2018).

NIST (National Institute of Standards and Technology). 2015. Lack of Effective Timing Signals Could Hamper "Internet of Things" Development. Available at https://www.nist.gov/news-events/news/2015/03/lack-effective-timing-signals-could-hamper-internet-things-development (accessed March 15, 2018).

Pagay, V., M. Santiago, D. A. Sessoms, E. J. Huber, O. Vincent, A. Pharkya, T. N. Corso, A. N. Lakso, and A. D. Stroock. 2014. A microtensiometer capable of measuring water potentials below –10 MPa. *Lab on a Chip* 14(15):2806-2817.

Pasquetto, I., B. Randles, and C. Borgman. 2017. On the reuse of scientific data. *Data Science Journal* 16:8.

Popkin, G. 2017. Quantum computer simulates largest molecule yet, sparking hope of future drug discoveries. *Science.* Available at http://www.sciencemag.org/news/2017/09/quantum-computer-simulates-largest-molecule-yet-sparking-hope-future-drug-discoveries (accessed March 15, 2018).

Priest, C. 2018. How Machine Learning Can Help Marketers Measure Multi-Touch Attribution. Available at https://www.ama.org/partners/content/Pages/how-machine-learning-can-help-marketers-measure-multi-touch-attribution.aspx (accessed July 9, 2018).

Salfer, J., M. Endres, W. Lazarus, K. Minegishi, and B. Berning. 2017. Dairy Robotic Milking Systems—What Are the Economics? eXtension. Available at http://articles.extension.org/pages/73995/dairy-robotic-milking-systems-what-are-the-economics (accessed April 26, 2018).

Seagate. 2017. Data Age 2025. Available at https://www.seagate.com/www-content/our-story/trends/files/Seagate-WP-DataAge2025-March-2017.pdf (accessed March 15, 2018).

Sennar, K. 2017. AI in Agriculture—Present Applications and Impact. Techemergence. Available at https://www.techemergence.com/ai-agriculture-present-applications-impact (accessed March 15, 2018).

Shekhar, S., J. Colletti, F. Munoz-Arriola, L. Ramaswamy, C. Krintz, L. Varshney, and D. Richardson. 2017. *Intelligent Infrastructure for Smart Agriculture: An Integrated Food, Energy and Water System.* Computing Community Consortium white paper. Available at https://arxiv.org/ftp/arxiv/papers/1705/1705.01993.pdf (accessed March 15, 2018).

Smith, W. 2017. Uncertainty and Opportunity: Recent News Clips of Interest. Presentation, AgGateway 2017 Annual Conference: Efficiency-Opportunity-Profitability, November 6-9, 2017, San Diego, California.

Stephens, Z. D., S. Y. Lee, F. Faghri, R. H. Campbell, C. Zhai, M. J. Efron, R. Iyer, M. C. Schatz, S. Sinha, and G. E. Robinson. 2015. Big data: astronomical or genomical? *PLoS Biology* 13(7):e1002195.

Top 500. 2018. Available at https://www.top500.org/lists/2018/06 (accessed March 15, 2018).

Turin, L. 2002. A method for the calculation of odor character from molecular structure. *Journal of Theoretical Biology* 216(3):367-385.

UN (United Nations). 2014. A World That Counts. Available at http://www.undatarevolution.org/report (accessed March 15, 2018).

USDA-NIFA (U.S. Department of Agriculture's National Institute of Food and Agriculture). 2017a. Big Data Yields Big Opportunities in Agriculture. Blog. Available at https://www.usda.gov/media/blog/2017/07/17/big-data-yields-big-opportunities-agriculture (accessed March 15, 2018).

USDA-NIFA. 2017b. Executive Summary: Changing the Face, Place, and Space of Agriculture. Available at https://nifa.usda.gov/sites/default/files/resource/Data%20Summit%20Summary.pdf (accessed March 15, 2018).

Vasisht, D., Z. Kapetanoic, J. Won, X. Jin, R. Chandra, A. Kapoor, S. Sinha, M. Sudarshan, and S. Stratman. 2017. FarmBeats: An IoT Platform for Data-Driven Agriculture. 14th USENIX Sympostium on Networked Systems Design and Implementation, March 27-29, 2017. Boston, MA. Available at https://www.usenix.org/system/files/conference/nsdi17/nsdi17-vasisht.pdf (accessed March 15, 2018).

Verdow, C., S. Wolfert, and B. Tekinerdogan. 2016. Internet of Things in Agriculture. Available at https://www.researchgate.net/publication/312164156_Internet_of_Things_in_agriculture (accessed March 15, 2018).

Wasik, B. 2013. In the Programmable World, All Our Objects Will Act as One. Available at https://www.wired.com/2013/05/internet-of-things-2 (accessed July 9, 2018).

Wilkinson, M. D., M. Dumontier, I. J. Aalbersberg, G. Appleton, M. Axton, A. Baak, N. Blomberg, J.-W. Boiten, L. Bonino da Silva Santos, P. E. Bourne, J. Bouwman, A. J. Brookes, T. Clark, M. Crosas, I. Dillo, O. Dumon, S. Edmunds, C. T. Evelo, R. Finkers, A. Gonzalez-Beltran, A. J. G. Gray, P. Groth, C. Goble, J. S. Grethe, J. Heringa, P. A. C 't Hoen, R. Hooft, T. Kuhn, R. Kok, J. Kok, S. J. Lusher, M. E. Martone, A. Mons, A. L. Packer, B. Persson, P. Rocca-Serra, M. Roos, R. van Schaik, S.-A. Sansone, E. Schultes, T. Sengstag, T. Slater, G. Strawn, M. A. Swertz, M. Thompson, J. van der Lei, E. van Mulligen, J. Velterop, A. Waagmeester, P. Wittenburg, K. Wolstencroft, J. Zhao, and B. Mons. 2016. The FAIR Guiding Principles for scientific data management and stewardship. *Scientific Data* 3:160018.

Wolfert, S., L. Ge, C. Verdouw, and M. Bogaardt. 2017. Big Data in smart framing—A review. *Agricultural Systems* 153:69-80.

Woodward, J. 2016. Big data and ag-analytics: An open source, open data platform for agricultural and environmental finance, insurance, and risk. *Agricultural Finance Review* 76(1):15-26.

Yang, Z., G. Yu, L. Di, B. Zhang, W. Han, and R. Mueller. 2013. Web service-based vegetation condition monitoring system—VegScape. Pp. 3638-3641 in *2013 IEEE International Geoscience and Remote Sensing Symposium (IGARSS)*. Institute of Electrical and Electronics Engineers.

Yli-Huumo, J., D. Jo, and S. Choi. 2016. Where is current research on blockchain technology? A systematic review. *PLoS ONE*. Available at https://doi.org/10.1371/journal.pone.0163477 (accessed March 15, 2018).

8

A Systems Approach

1. INTRODUCTION

The food system is a complex system with a variety of actors, feedback and interdependence mechanisms, heterogeneity, and spatial and dynamic complexity (IOM and NRC, 2015). The system is composed of large-scale interconnected systems of systems that include ecosystems (both natural and agricultural), climate, food processing and distribution network data and information systems, and socioeconomic systems. Models are used to better understand how various entities interact within and integrate across the system, and to understand relationships among the natural environment, society, economy, and the various biophysical components of the food system. For example, an understanding of those relationships enables agroecology and ecohydrology approaches to be used to simultaneously manage multiple goals (e.g., lowering water use, sequestering carbon, and minimizing environmental impacts). However, modeling complex interactions and feedback loops within and between each of these systems can be improved through (1) a better understanding of the drivers (e.g., physical, behavioral, market drivers) and their interactions, and (2) a greater ability to collect and analyze large amounts of data with high geospatial and temporal variability.

The study of such a complex system will serve the ultimate goal of informing decision making in food and agriculture practices and policies. Such systems models are also needed to evaluate the impacts of a technological breakthrough on the sustainability of the food system, to rapidly respond to perturbations in the systems to avoid adverse outcomes, and to propose interventions that lead to overall efficiency gains. The socioeconomic elements of the system need to be deeply integrated with the biophysical drivers of the agroecosystem to invoke any meaningful changes. This chapter presents the many challenges in conducting systems-level models and analysis of the food system, and the opportunities to advance this scientific field.

2. CHALLENGES

2.1 Challenge 1: Create and Link Efficient Integrated Systems Models to Data Collection Efforts

In describing subsystems of the food system, models have been successfully built that are biophysics based (based on understanding of the behavior of a system's components) and empirically based (based on observation and measurements). For example, soil and water balance models were developed to understand climate constraints on agriculture, soil models were developed to predict impacts of soil erosion on crop productivity, and insect and disease models were developed for integrated pest management strategies (Jones et al., 2017). In soils, great strides have been made in understanding N_2 fixation and uptake and in determining types of perturbations that may impact the N_2 cycle (Salvagiotti et al., 2008). Models for predicting weather at hyperregional scales are available, such as the National Center for Atmospheric Research's Weather Research and Forecasting model and The Weather Company's Deep Thunder (The Weather Company, 2016; NCAR, 2018). However, these various modeling efforts are generally discipline-specific and the subsystem models are not yet well integrated across all of the different disciplines within the food and agricultural system. There are many challenges to creating a reliable integrated systems model for the food system.

Systems Are Not Well Defined

The systems are not well defined, and in many cases the linkages between the various systems are not delineated well enough to support mechanistic-based physical models. For example, how do changes in the microbiome of soil microorganisms related to N_2 fixation affect the resilience of plants against heat or salt stress, to the nutritional value of the food produced, or to feed conversion efficiency in animals? How would a mechanistic understanding of the interactions of the soil microbiome with the plant, root, and leaf microbiomes be combined with genetic tools to create more resilient plants?

In addition to linkages, a significant challenge in defining any particular subsystem is determining the main drivers of change for consideration in the analysis. For example, to help identify means to address the differences in productivity and other outcomes between and within regions, it would be helpful to measure and analyze the causes (e.g., drivers) for such differences. In another example, although trade-off analysis is a useful tool for integrating biophysical models with socioeconomic models to predict the impacts of different drivers on the oversustainability of the system (Stoorvogel et al., 2004), these models require a priori assumptions about the principal drivers affecting the model outcomes, and consider only a subset of all possible drivers. Life-cycle analysis (LCA) takes a more holistic approach to modeling sustainability indicators, and can be useful for determining, for instance, the range of the magnitude of life-cycle impacts for a given agriculture activity (e.g., milk production) or set of dietary food choices. This can be used to identify system-level opportunities for intervention to improve sustainability (Boehm et al., 2018; Poore and Nemecek, 2018). However, the outcome of the LCA will depend on whether it is being optimized for water use, yield, or nutritional content (Notarnicola et al., 2017; Pourzahedi et al., 2018). The outcome may also depend on the system boundaries that are selected.

Lack of Cohesion Between the Different Modeling Communities

As previously noted, the absence of standardization of data formats and reporting, data privacy concerns, proprietary data, and lack of data repositories are impediments to data sharing. Data standardization and open access are important since the output of one system model will likely be the input to a different system model.

Lack of Methods for Integrating Models

Better methods for integrating subsystem models across different spatial and temporal scales and for integrating physics-based mechanistic models with empirical/statistical models or economic models are needed (Kling et al., 2017). Mechanistic (biophysical) models of the key components of a single system may be possible, but adding only a few additional nodes to a system analysis creates significantly more connections that may each take many years of research to achieve (see Box 8-1). This precludes the ability to create biophysics-based mechanistic models for the food system. The problem of scale is exacerbated by the fact that many different forms of perturbations can affect the systems simultaneously (e.g., society's values and climate). Each of the system nodes has complicated feedback loops that will need to be considered together in a systems-level analysis. Tackling this challenge will likely require the development of empirical (statistical)

BOX 8-1
Systems Aspects of the Food System

Scale is a complicating factor in systems analysis of the food system (see Figure 8-1). Significant research has aimed at understanding the biology, chemistry, and physics to create and validate biophysics-based *mechanistic* models of selected subsystems. These are typically small-scale models including only the key drivers of the system. There are also models of large-scale systems, for example, global increases in temperature or precipitation due to increased CO_2 in the atmosphere. These models are physics based but often semi-empirical or data driven, rather than purely (first principles) mechanistic, and require averaging over large areas, simplifying assumptions, and statistics. The research community needs to effectively bridge these scales. However, coupling systems having only a few additional elements create the need to understand significantly more causal relationships, each of which make take many years of research to achieve. The problem of linking process across scales is not unique to the food system.

Overarching questions include

- How to bridge across scales, including spatial-temporal scales as well as with the number of processes included in the "system" being studied; and
- How to link mechanistic models developed for smaller, better-defined subsystems with large-scale statistical models used to predict macroscopic behaviors such as global carbon storage in soils.

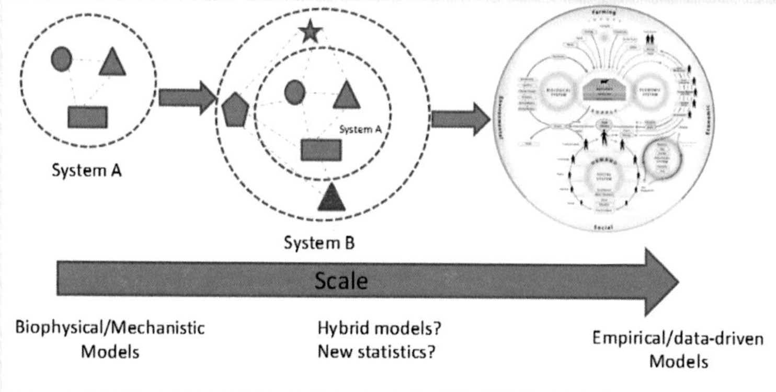

FIGURE 8-1 The scale used to model a system is dependent on the number of components considered. Each additional node in a system greatly increases the number of interactions that need to be elucidated for a systems model. The food and agricultural system has a large number of nodes, and thus corresponding interactions, which makes the system complex.
SOURCE: Food System Map courtesy of Nourish Initiative, www.nourishlife.org. Copyright WorldLink, all rights reserved.

models, rather than purely mechanistic ones. The spatial resolution of many components of the food system may be quite coarse (e.g., soil moisture determined from satellite sensing has kilometer to tens of kilometers scale resolution), which will require averaging, simplifying assumptions, and conducting statistics to downscale the models for more-local predictions, including analysis of differences within and between regions. However, better ground sensing as described in Chapter 3 may improve this situation. Methods to blend these two types (and scales) of models (mechanistic and empirical) and data are needed.

Updating Models

Sensor development and movement toward data-driven empirical models for the food system will require massive amounts of high-quality data to calibrate the models. The current approaches for integrating data into models to update model parameters are slow. Rapid (perhaps real-time) updating of the models with data is necessary, whether the subject is water and nutrient use in crop agriculture, feed efficiency in animal agriculture, or food safety and quality in food processing and distribution.

2.2 Challenge 2: Uncertainty Within and Across Multiple Elements of the System

The high degree of uncertainty in the food system is a significant challenge for developing accurate models. Uncertainty is especially challenging for long-range strategic planning efforts to cope with a growing population. Uncertainty comes in many forms; for example, it is difficult to predict extremes in environmental conditions (water, heat), yields, pricing, and market conditions. While it is also difficult to predict the behavior of growers and consumers related to technology adoption and new inputs, it is important for the models to take behavior into account. Incorporating information on adoption behaviors and market response is required to best quantify response to the effects of technology changes, and useful examples include recent applications to water (Khanna and Zilberman, 2017; Taylor and Zilberman, 2017; Zilberman et al., 2017). There is a lack of systems-level efficiency data (e.g., water-use or nitrogen-use efficiency at the watershed scale) available to calibrate the models. High degrees of uncertainty across multiple system elements can make accurate predictions of overall system behavior inherently indeterminate.

2.3 Challenge 3: Access to Agricultural Systems Models and Better Integration of Management and Society into the GEMS Model

Agricultural systems modeling is a well-established field. Significant progress has been made in understanding the biophysical processes driving agricultural productivity (Jones et al., 2017), and integrating biophysical and economic models for analysis (van Wijk et al., 2014). To date, modeling efforts have focused on understanding the linkages between system components and drivers of the system responses. Challenges remain in making these models more accessible to users, in creating models that can address specific user needs (e.g., including interactions between multiple crops and between crops and livestock), and in developing user-friendly decision tools (Antle et al., 2017). Another challenge in agricultural systems modeling is in data management. There are vast amounts of farm-level data available, and effectively storing, validating, and sharing those data are crucial to developing the next generation of agricultural systems-level models.

The genotype-environment-management-society (GEMS) approach to modeling agricultural systems offers great opportunities to improve system efficiency by acknowledging complex interactions. Expanding the existing systems models to include additional societal drivers would enable better predictions of behavior in the food system, and is necessary to invoke meaningful changes to the system. For example, food choices and behaviors affecting food waste could have significant impacts on the overall sustainability of agriculture. The price of commodities is driven by policies (e.g., biofuel mandates), technology choice, and consumer demands, which in turn affects the mix of crops selected by growers. These do not always align with climate or resource availability if prices and other factors do not fully reflect the opportunity cost of alternative uses.

3. OPPORTUNITIES

This section includes three opportunities to advance the systems approach for food and agriculture. Breakthroughs in agricultural research will come from the collaborative efforts of scientists and experts in fundamental and applied science areas, policy, and human behavior. An example of systems thinking and integrated systems modeling for watershed management is noted in Box 8-2.

BOX 8-2
Example: Management Solutions to Decrease Hypoxic Zones

The formation of hypoxic zones in drainage basins provides a good example of where systems thinking and integrated systems modeling can be used to identify optimal (lowest-cost) management solutions. Rabotyagov et al. (2014) developed an integrated assessment model to determine the impact of different cropland conservation investment decisions in more than 550 agricultural subwatersheds that deliver nutrients into the Gulf of Mexico on the aerial extent of hypoxia. By considering the whole system simultaneously, the location, type, and extent of interventions needed to achieve a desired reduction in aerial extent of hypoxia in the Gulf could be estimated. The optimal locations of intervention and the trade-offs between cost and outcome were determined (see Figure 8-2). The approach is extensible and can include other outcomes, for example, ecosystem services, impacts on water quality, and soil conservation. This enables creation of a multidimensional trade-off frontier that could be used to make more informed policy decisions.

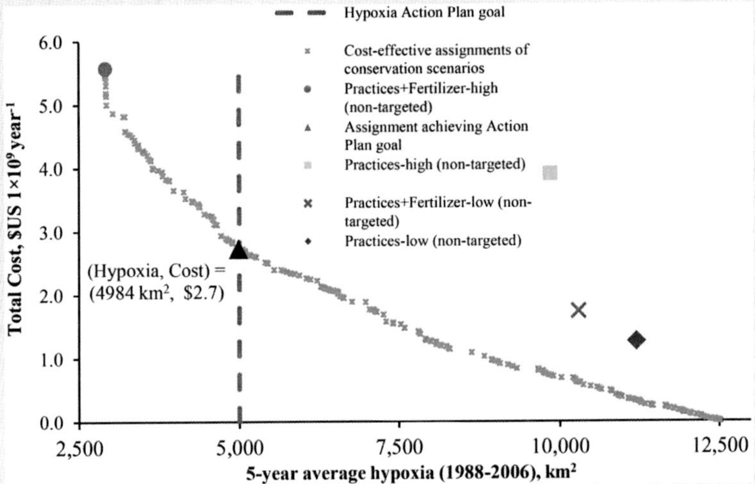

FIGURE 8-2 The trade-offs of cost and hypoxia by various cropland conservation scenarios across subwatersheds. Selecting the most cost-effective assignments for conservation in each subwatershed provides the lowest cost to achieve the state goal of keeping the hypoxic zone under a total area of 5,000 km^2.
SOURCE: Rabotyagov et al., 2014.

3.1 Opportunity 1: Complex System Modeling to Determine Methods for Integrating Data, Models, Management Strategies, and Socioeconomic Behaviors

Development of data-driven integrated food system models will inform future research efforts toward the most promising agricultural breakthrough opportunities to be pursued, or the best management practices for efficient resource use. The process of continuously updating and validating models with data will likely lead to more mechanistically based data-driven models and advancements in minimizing and managing uncertainty within and between systems to improve forecast reliability. Matching model complexity to scale and to user needs can result in better decision support tools for producers.

As previously mentioned, the study of the food and agricultural system as a complex system serves the ultimate goal of informing decision making and—specifically relevant to this report—of evaluating the impacts of technological breakthroughs on the sustainability of the food system. The complexity of the food system, the integration of its various systems, including the natural and built environments, and the significant need to increase its resilience and efficiency make the food system and its subsystems an ideal model for developing the capacity to model complex systems in general. Initiatives to study the food system provide the opportunity to determine appropriate boundaries for analysis within and between systems, to identify the key data gaps and missing linkages within systems and at the boundaries of different systems, and to identify and build the required systems analysis and decision support tools for complex agricultural systems. There is an opportunity to identify appropriate "behavior" models to include in agricultural systems models, and to standardize different system models through model intercomparisons using "base-case" scenarios for model comparison. As weather and climate models increase resolution and better represent small-scale processes such as vegetative feedbacks, it will open research into soil–plant–atmosphere–water cycling from field to continental scales. Modeling the system response to climate extreme events can improve the utility of the models to growers. Scaling up models from field scale to landscape scale can also help to manage agroecosystems coupled with natural ecosystems using ecohydrology methods.

3.2 Opportunity 2: Pioneer New Methods for Improving Data-Sharing and Privacy Issues for Public–Private Partnerships

As previously noted in Chapter 7, data-driven approaches to managing agriculture offer unprecedented opportunity for increasing the sustainability of the food system. The availability of open, harmonized data has already proven to be essential to creating models of the food system (Jones et al., 2017). Continued development and applications of these approaches to the food system provides an opportunity to discover new methods of data collection, provenance, and sharing. It also provides opportunities to integrate data into models in real time, and to develop machine-learning and artificial intelligence approaches for managing inputs and outputs along the food value chain. There is a great opportunity to advance systems modeling by developing methods to integrate both public and private data for use in democratized decision support tools (Antle et al., 2017).

3.3 Opportunity 3: Apply Systems Thinking and Convergence Methods to Agricultural Sustainability Problems

The food system touches on a wide range of issues, including climate change, water resource use, plant and animal genetics, food safety, nutrition, policy, and economic sustainability. The focus on individual subsystems, without consideration of the interconnectedness of the subsystems, has led to unintended consequences to other components of the system (Turner et al., 2016). In other words, approaching agricultural research more holistically from a system-of-systems perspective offers the potential to optimize across multiple components of the system. The basic tenet of "systems thinking" is recognizing the interconnected nature and linkages that cross multiple dimensions (subsystems). There is also tremendous opportunity to promote convergence of knowledge to solve agriculture's most vexing problems (e.g., water use, soil degradation, and food waste). For example, combining expertise from plant biology, microbiology, soil science, nanotechnology, sensor design, wireless communications, and behavioral sciences may lead to innovative and socially acceptable methods of applying nanotechnology-based sensors into plants and soils to monitor plant productivity without negatively affecting the plants, environment, or the ability to sell the food produced. Combining expertise may help to convert "waste" products to value-added products or to important sources of inputs and energy to other parts of the food system. Programs aimed at tackling all of the elements of these problems will require better integration of fundamental science and engineering with economics, medicine, behavioral sciences, and social sciences. New knowledge and research is needed on how to design technology packages and institutional structures to enhance technology adoption and new processes. Research in systems-thinking approaches to agricultural problems will ultimately trickle down into new educational programs and into practice as those graduates enter the profession.

4. BARRIERS TO SUCCESS

Barrier 1: Data provenance, security, sharing, and costs of collection and storage. Data-driven approaches to manage the food system offer great potential to improve resilience and sustainability. However, issues around governance and the development of a viable business model for democratizing and sharing the benefits derived from those data are significant barriers to success (Halewood et al., 2018). Institutional and policy challenges are needed to promote the formation of public–private partnerships, determine ownership of resources, and manage sharing of benefits across countries. The lack of appropriate platforms, standards, and incentives for sharing of data and models is also a substantial barrier to success.

Barrier 2: Systems are inherently indeterminate. There need to be realistic expectations about the applicability and accuracy of the food system models. As Box 8-1 showed, as more subsystems are included in the analysis, the number of subsystem interactions increases considerably. The causal relationships between many elements of the food system are not well characterized. Moreover, the high spatial and temporal variability within these systems and between the systems and the environment, as well as the unpredictability of human behaviors, may simply make the system too complex to model with reasonable accuracy. As an example, the predominant methods for irrigation in water-limited western states are flood and furrow irrigation systems. Although these are highly inefficient systems, farmers fear that switching to more efficient sprinkler and drip irrigation systems will erode their stake in and ownership to their water rights (i.e., the "use it or lose it" syndrome), which is likely the reason many farmers hold back from making changes to more efficient irrigation systems (Osborn et al., 2017). Evaluating the effects of policy-induced changes to incentives for switching to the more efficient irrigation systems requires the development of coupled models that account both for the geospatial component and water usage as well as for human behaviors related to adaptation and induced technological change.

Barrier 3: Limited incentives to consider systems-level impacts. Policies and regulations affecting agriculture (e.g., water policy, water quality criteria, and nutrient management) are often made without consideration of their impact on the system (Alfredo and Russo, 2017). The research enterprise has not been structured with the goal of applying systems approaches to problem solving across the span of the food and agricultural system, and there are limited incentives for research cooperation among disciplines.

Barrier 4: Unintended consequences are not immediately identifiable. Unidentified risks in any system or system of systems are unavoidable. The more complex and interrelated the system becomes, the greater the potential for unidentified risks. Although it is impossible to identify all of the risks of any system, a move toward systems thinking in food and agricultural research will help to reduce any unintended consequences due to interventions—for example, altering photosynthesis routes, manipulation of the soil microbiome, or environmental costs of information and communication technologies associated with widescale deployment of autonomous data-collecting sensors.

5. RECOMMENDATIONS

Recommendation 1: Identify opportunities to improve the performance and adoption of integrated systems models of the food system and decision support tools. The community of stakeholders should identify current barriers to successfully applying integrated systems models to the food system. The community should also develop a roadmap to overcome key challenges specific for modeling the food system (e.g., integration of subsystems, standardization and interoperability, data storage and sharing, data infrastructure, simplified decision support tools, crop–crop and crop–livestock interactions, and validating the performance of complex systems models using data). The community should clearly delineate where the inability to model the integrated system components is limiting progress toward sustainability and resilience, rather than where other barriers exist such as poor policies or undervaluation of natural resources.

Once the key systems and system of systems to be modeled are determined, the research community needs to improve understanding of the drivers behind the subsystem interactions. This will require improving the ability to collect and analyze large amounts of data with high geospatial and temporal variability (e.g., sensor development). Connecting the submodels into more integrated systems models will also require consistency in applying system boundaries and assumptions and scale of analysis (Kling et al., 2016), as well as more standardized data reporting and model input/output to make the subsystems interoperable. These integrated system models can then be used to direct research needs and to avoid unwanted consequences of technology interventions or policies.

Recommendation 2: Incorporate elements of systems thinking and sustainability into all aspects of the food system (from education to research to policy). A paradigm shift is needed in the management of the food system to anticipate the effects of environmental or policy-induced changes. This will be a fundamental change in the way the food system is viewed, and the way stakeholders are educated to operate. In research, it will require more transdisciplinary research and team science around approaching and working to solve problems in the food system. It will also require more integrative, systems-level approaches that can assess policy alternatives for systems-level sustainability outcomes. It will also require adequate incentives and funding for systems-level convergent approaches to research and education.

REFERENCES

Alfredo, K. A., and T. A. Russo. 2017. Urban, agricultural, and environmental protection practices for sustainable water quality. *WIREs: Water* 4(5). doi: 10.1002/wat2.1229.

Antle, J. M., B. Basso, R. T. Conant, H. C. J. Godfray, J. W. Jones, M. Herrero, R. E. Howitt, B. A. Keating, R. Munoz-Carpena, C. Rosenzweig, P. Tittonell, and T. R. Wheeler. 2017. Towards a new generation of agricultural system data, models and knowledge products: Design and improvement. *Agricultural Systems* 155:255-268.

Boehm, R., P. E. Wilde, M. Ver Ploeg, C. Costello, and S. B. Cash. 2018. A comprehensive life cycle assessment of greenhouse gas emissions from U.S. household food choices. *Elsevier* 79(C):67-76.

Halewood, M., T. Chiurugwi, R. S. Hamilton, B. Kurtz, E. Marden, E. Welch, F. Michiels, J. Mozafari, M. Sabran, N. Patron, P. Kersey, R. Bastow, S. Dorius, S. Dias, S. McCouch, and W. Powell. 2018. Plant genetic resources for food and agriculture: Opportunities and challenges emerging from the science and information technology revolution. *New Phytologist* 217(4):1407-1419.

IOM and NRC (Institute of Medicine and National Research Council). 2015. *A Framework for Assessing Effects of the Food System*. Washington, DC: The National Academies Press.

Jones, J. W., J. M. Antle, B. Basso, K. J. Boote, R. T. Conant, I. Foster, H. C. J. Godfray, M. Herrero, R. E. Howitt, S. Janssen, B. Keating, R. Muñoz-Carpena, C. Porter, C. Rosenzweig, and T. R. Wheeler. 2017. Brief history of agricultural systems modeling. *Agricultural Systems* 155:240-254.

Khanna, M., and D. Zilberman. 2017. Inducing water conservation in agriculture: Institutional and behavioral drivers. *Choices* 32(4).

Kling, C. L., R. W. Arritt, G. Calhoun, and D. A. Keiser. 2017. Integrated assessment models of the food, energy, and water nexus: A review and an outline of research needs. *Annual Review of Resource Economics* 9:143-163.

NCAR (National Center for Atmospheric Research). 2018. Weather Research and Forecasting Model. Available at https://www.mmm.ucar.edu/weather-research-and-forecasting-model (accessed June 8, 2018).

Notarnicola, B., S. Sala, A. Anton, S. J. McLaren, E. Saouter, and U. Sonesson. 2017. The role of life cycle assessment in supporting sustainable agri-food systems: A review of the challenges. *Journal of Cleaner Production* 140:399-409.

Osborn, B., A. S. Orlando, D. L. Hoag, T. K. Gates, and J. C. Valliant. 2017. The Economics of Irrigation in Colorado's Lower Arkansas River Valley. Available at http://www.cwi.colostate.edu/media/publications/sr/32.pdf (accessed June 26, 2018).

Poore, J., and T. Nemecek. 2018. Reducing food's environmental impacts through producers and consumers. *Science* 360(6392):987-998.

Pourzahedi, L., M. Pandorf, D. Ravikumar, J. Zimmerman, T. Seager, T. Theis, P. Westerhoff, L. Gilbertson, and G. V. Lowry. 2018. Life cycle considerations of nano-enabled agrochemicals: Are today's tools up to the task? *Environmental Science: Nano* 5:1057-1069.

Rabotyagov, S. S., T. D. Campbell, M. White, J. G. Arnold, J. Atwood, M. L. Norfleet, C. L. Kling, P. W. Gassman, A. Valcu, J. Richardson, R. E. Turner, and N. N. Rabalais. 2014. Cost-effective targeting of conservation investments to reduce the northern Gulf of Mexico hypoxic zone. *Proceedings of the National Academy of Sciences of the United States of America* 111(52):18530-18535.

Salvagiotti, F., K. G. Cassman, J. E. Specht, D. T. Walters, A. Weiss, and A. Dobermann. 2008. Nitrogen uptake, fixation and response to fertilizer N in soybeans: A review. *Field Crops Research* 108(1):1-13.

Stoorvogel, J. J., J. M. Antle, C. Crissman, and W. Bowen. 2004. The tradeoff analysis model: Integrated biophysical and economic modeling of agricultural production systems. *Agricultural Systems* 80(1):43-66.

Taylor, R., and D. Zilberman. 2017. Diffusion of drip irrigation: The case of California. *Applied Economic Perspectives and Policy* 39(1):16-40.

Turner, B. L., H. M. Menendez, R. Gates, L. O. Tedeschi, and A. S. Atzori. 2016. System dynamics modeling for agricultural and natural resource management issues: Review of some past cases and forecasting future roles. *Resources* 5(4):40.

van Wijk, M., M. Rufino, D. Enahoro, D. Parsons, S. Silvestri, R. Valdivia, and M. Herrero. 2014. Farm household models to analyse food security in a changing climate: A review. *Global Food Security* 3:77-84.

The Weather Company. 2016. The Weather Company Launches "Deep Thunder"—the World's Most Advanced Hyper-Local Weather Forecasting Model for Businesses. Press Release. Available at http://www.theweathercompany.com/DeepThunder (accessed June 26, 2018).

Zilberman, D., R. Taylor, M. E. Shim, and B. Gordon. 2017. How politics and economics affect irrigation and conservation. *Choices* 32(4).

9

Strategy for 2030

1. INTRODUCTION

Remarkable advances in food and agricultural sciences over the past century demonstrate how public support for research can enable talented U.S. scientists to contribute to improving food and agriculture (NRC, 2014b). Much has been accomplished, yet much remains to be done. Food and agricultural research will need to respond to and address some of the looming contingencies that threaten the resiliency of our food system. The United States is facing resource scarcities in some regions (such as water shortages in the High Plains), increased variability in weather conditions, and other rapid transformations in the agricultural sector that may be difficult to reverse. For some crops (such as soybeans), it is no longer possible to continue increasing inputs to maximize yields or maintain nutritional value without sacrificing other resources (such as the contamination and depletion of water and soil). In animal agriculture, imported animals and animal products endanger our national herds and flocks through the potential introduction of diseases that would decimate our resources (such as avian influenza and foot-and-mouth disease). Further innovations in management are needed to reduce waste and greenhouse gas emissions per animal. Food loss and waste remains at an estimated 30-40 percent, with key resources lost primarily at the retail and consumer stages (Buzby et al., 2014). Therefore, there is a need to find new methods, processes, and systems to better handle and preserve our food supply and to optimize and better utilize existing ones. Diversifying our approaches will be key to strengthening and improving the resiliency of our food supply.

In the past decade, the United States has lost its status as the top global performer of public agricultural research and development (R&D) to China (Clancy et al., 2016). Unless the United States reverses this trend and invests, the most promising innovations to enhance productivity and expand the food supply are likely to come from other countries, and the United States will fall behind other countries in terms of agricultural growth. There is an urgent need to facilitate the convergence of disciplines and science breakthroughs to address the "wicked" problems[1] facing U.S. food and agriculture. Tackling these challenges will require a leveraging of capabilities across the scientific and technological enterprise. In addition, talent from allied disciplines, such as behavioral and economic sciences, will be indispensable to help solve these problems. In some cases, new tools (e.g., CRISPR) are available, but knowledge is still lacking as to how to effectively deploy those tools (e.g., in developing more resilient crops and better food quality and nutritional characteristics) and how to effectively gain producer and consumer acceptance. In other cases, it appears that there are promising opportunities in allied fields such as data science and sensors, but it remains unclear how researchers from these disparate disciplines will be enticed to collaborate on research programs to provide solutions that benefit food and agriculture. For instance, data will be essential for accurate predictions or assessments, and the opportunities appear endless in how big data and analytics can be used to increase efficiencies across the food and agricultural spectrum (e.g., to optimize breeding and improve food handling and distribution).

Food and agricultural research will play a critical role in meeting the food needs in the United States and elsewhere in a sustainable manner in the upcoming decade. Now is the time to develop a strategy that would harness emerging scientific advances to transform agricultural systems for greater efficiency, resiliency, and sustainability. Given the generational time from bench to field, investments are needed today to provide future benefits. Much can be accomplished in 10 years.

This report proposes a strategic path forward by first examining the solutions to problems through the concept of convergence and then by providing recommendations for carrying out such a strategy through 2030. For the United States to maintain its global leadership in food and agricultural

[1]A "wicked" problem is an intractable problem with many interdependent factors that make it difficult to define and solve (Rittel and Webber, 1973; Crowley and Head, 2017). The term "wicked" is used not to indicate that a problem is ethically deplorable, but it instead refers to the challenging nature of the problem. Wicked problems are too complex to be solved using conventional linear approaches, and instead require rationale systematic processes for solutions. The concept was first introduced by Rittel and Webber (1973) to address social concerns and has since been widely applied to address problems with a social component such as environmental degradation and climate change.

research, it will be critical to invest both public and private resources to achieve scientific breakthroughs and to develop the critical scientific workforce that can address the multifaceted challenges. A better understanding of the food system itself can help better manage the overall system and its impacts.

2. CONVERGENCE

The food system is vast and complex, making it challenging to comprehensively address all of the issues included in food production. In the past, it has been more common to examine these problems in a defined space or discipline for reasons related to practicality and greater ease of management, and that approach has been effective at addressing distinct issues that require specific knowledge in a domain. However, the so-called wicked problems of the future will require a radically different approach to understand the issues and uncover solutions that can only be found when explored beyond the traditional boundaries of food and agricultural disciplines.

Convergence is described as

> an approach to problem solving that cuts across disciplinary boundaries [and] integrates knowledge, tools, and ways of thinking from life and health sciences, physical, mathematical, and computational sciences, engineering disciplines, and beyond to form a comprehensive synthetic framework for tackling scientific and societal challenges that exist at the interfaces of multiple fields. (NRC, 2014a; p. 1)

The National Science Foundation (NSF) has identified convergence research as one of the 10 big ideas for future investments (NSF, 2018). Food and agricultural research would greatly benefit from incorporating convergent approaches, as convergence merges diverse areas of expertise to stimulate innovation ranging from basic science discoveries to translational scientific applications (NRC, 2014a).

The next decade will be a critical juncture for food and agricultural research. For instance, the emergence of a digital world has given rise to the ability to collect, share, and analyze information much more easily and readily than in the past. It will be important to harness the opportunities that data have to offer in order to integrate and harmonize data so that they are usable and accessible. Boxes 9-1 and 9-2 provide examples of specific problems that will require convergence to address the various issues within the broader system.

BOX 9-1
Water Usage

Crop agriculture in arid regions that are being stressed further by increased climate variability and dwindling water resources provides an excellent example of where the convergence of disciplines discussed in this report could increase the sustainability and resilience of agriculture. Current water consumption for crop agriculture in the Central High Plains Region of the United States is not sustainable due in part to groundwater depletion in the Ogallala Aquifer (Steward et al., 2013). Applying the breakthroughs and technological advances described in this report in an integrated systems-oriented way will offer unprecedented opportunities to address water use and sustainability of crop agriculture in these types of arid regions. Advances in sensors (ground and remote) and data analytics, better seasonal weather forecasts at high spatial and temporal resolution, and enabling the safe use of alternative water sources for agriculture (e.g., wastewater) can be combined to enable data-driven automated water management that maximizes the productivity of the water sources that are available. Achieving such data-driven approaches will also require faster computing, data storage, quality control, democratizing of data, and the ability to use data in real time to update and improve the accuracy of physics-based models describing water use in this agricultural system. Coupling these advances in water management with new genetic tools that can provide a diverse collection of drought- and disease-tolerant plants, improved photosynthetic and water-use efficiency, coupled with a soil/phytomicrobiome that enhances water and nutrient uptake, can lead to more sustainable crops in arid regions. Decreasing food loss and waste (which means that fewer water inputs will be lost overall) and changing the policies and pricing of water are also needed. No single breakthrough is likely to achieve sustainability

3. RECOMMENDATIONS FOR STRATEGY

3.1 Major Goals

To achieve the major goals of efficiency, resiliency, and sustainability, improvements are needed to address the most challenging issues across the food system. The key research challenges identified by the committee include (1) increasing nutrient use efficiency in crop production systems, (2) reducing soil loss and degradation, (3) mobilizing genetic diversity for crop improvement, (4) optimizing water use in agriculture, (5) improving food animal genetics, (6) developing precision livestock production systems, (7) early and rapid detection and prevention of plant and animal diseases, (8) early and rapid detection of foodborne pathogens, and (9) reducing food loss and waste throughout the supply chain.

with respect to water use but combining all of these practices will have significant impact. A similar approach could be applied to address problems in other regions (e.g., flooding). Figure 9-1 depicts how water availability and consumption issues require a convergent approach in finding multiple types of solutions.

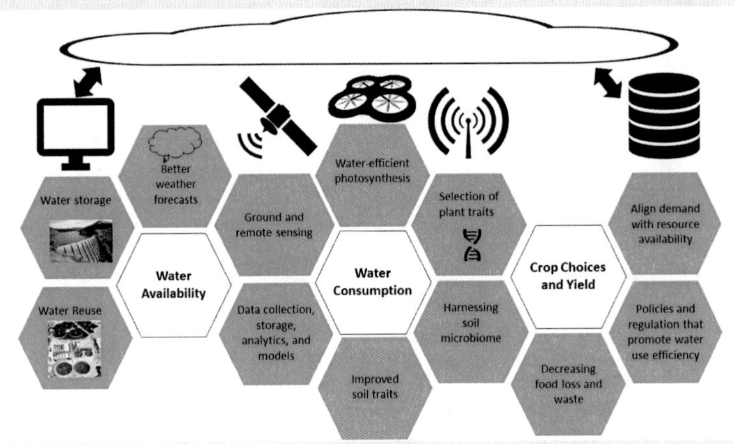

FIGURE 9-1 Convergent approaches for addressing water use in agriculture. Increased water-use efficiency and productivity can be achieved through data-driven, automated water management, more resilient and efficient crops, better seasonal weather forecasts, and policies and market forces to align demand with resource availability.

In addressing these challenges, the committee identified important research directions by disciplines or categories that encompass the overarching and interrelated goals of efficiency, resiliency, and sustainability. The recommended research directions in the preceding chapters are important opportunities because of new scientific developments that make them possible in the near term.

3.2 Science Breakthroughs and Recommendations

Solving the most vexing problems in the next decade will require a strategy that promotes convergent problem-solving approaches and scientific and technological breakthroughs that span across disciplines. The previous chapters identified challenges and opportunities by research disciplines and recommended specific research directions in those fields. Five

BOX 9-2
Animal Disease and Food Safety

Developing information for an integrated systems approach helps support more effective animal disease approaches and food safety outcomes. Bovine respiratory disease (BRD) is the leading cause of morbidity and mortality in both dairy and beef cattle and is caused by a variety of viruses and bacteria. It can greatly hamper an animal's productivity and welfare, may result in death, and is estimated to cost the U.S. cattle industry approximately $3 billion annually. Despite the fact that BRD has been extensively studied since the 1800s, it remains prevalent. Different sectors of the supply chain are disparately affected by disease incidence, yet currently no data are shared. For example, a cow-calf producer might vaccinate a calf prior to sale, but that information is often not passed on to the feedlot sector, and therefore all calves are revaccinated upon feedlot entry with additional costs to producers. This disincentivizes vaccination on the ranch and may result in many animals being needlessly vaccinated twice, with little return for the second vaccination. Developing technologies and information systems offer a new approach to tackle BRD by integrating the information from the different segments of the supply chain (such as the use of blockchain). This could offer valuable information that was previously unavailable to make genetic improvement in selection for animals that are less susceptible to disease. It could also provide information to breeders to allow accelerated improvement toward selection for both product quality and health, manage animals based on sensor and weather inputs and predictions, and trace both animal product quality and quantity for the retail chain. Improved trace-back systems also support more rapid and efficient product tracing in case of food safety recalls due to contamination in meat and meat products (e.g., for *E. coli* O157:H7). Collectively these breakthroughs will decrease animal disease incidence, thereby decreasing the use of antibiotics, and in combination will address multiple issues, including traceability for disease and food safety concerns, improved feeding and precision manage-

critical breakthrough opportunities emerged in the study that the committee found to be applicable across many different disciplines. Some of these are in the early stages of development, and others are on the cusp for widespread application to food and agriculture. The five areas identified as the most promising science breakthroughs for food and agricultural research are intended to facilitate the recommended research directions and serve the overarching goals of efficiency, resiliency, and sustainability. The recommendations that follow will require a shift in how the research community approaches its work, and initiatives for each of the breakthroughs will require robust support.

ment of animals, and traceability/track-back of animal products to inform genetic improvement programs, to ensure food safety, and to provide product information to consumers. Challenges remain with producers and others concerned about sharing data. Research on data technologies, behavioral incentives, and policy mechanisms can help to address these challenges. Figure 9-2 depicts how animal disease could benefit from a systems-based convergent approach in finding multiple types of solutions.

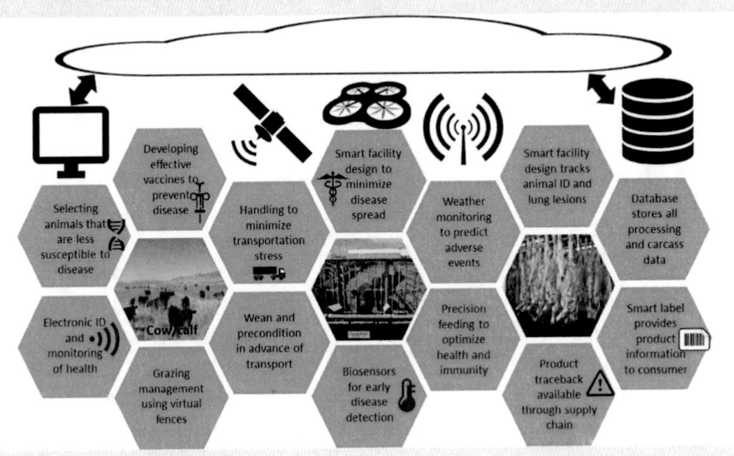

FIGURE 9-2 Convergent approaches to address animal disease and food safety. Precision livestock farming makes animal agriculture safer and healthier for animals, people, the environment, and the economy.

Transdisciplinary Research and Systems Approach

Breakthrough 1: A systems approach to understand the nature of interactions among the different elements of the food and agricultural system can be leveraged to increase overall system efficiency, resilience, and sustainability. Progress in meeting major goals is only able to occur when the scientific community begins to more methodically integrate science, technology, human behavior, economics, and policy into biophysical and empirical models. For example, there is the need to integrate the rate and determinants of adopting new technologies, practices, products, and processing innovations into food and agricultural system models. This approach is required to properly quantify the shifts in resource use, market effects, and response (including the effects of scale), as well as to determine benefits that are achievable from the scientific and technological breakthroughs (Khanna and Zilberman, 2017; Perry et al., 2017; Taylor and Zilberman, 2017). Considerations of these system interactions is critical for finding holistic solutions to the food and agricultural challenges that threaten our security and competitiveness.

Recommendation 1: Transdisciplinary science and systems approaches should be prioritized to solve agriculture's most vexing problems. Solving the most challenging problems in agriculture will require convergence and systems thinking to address the issues; in the absence of both, enduring solutions may not be achievable. Transdisciplinary problem-based collaboration (team science) will need to be facilitated because for some, it is difficult to professionally gravitate to scientific fields outside of their expertise. Such transitions will require learning to work in transdisciplinary teams. Enticing and enabling researchers from disparate disciplines to work effectively together on food and agricultural issues will require incentives in support of the collaboration. The use of convergent approaches will also facilitate new collaborations that may not have occurred when approached by researchers operating in disciplines in separate silos. Transdisciplinary problem-based collaborations will enable engagement of a new or diverse set of stakeholders and partners and benefit the food and agriculture sector (NRC, 2014a). Leadership is key to making team science successful, as scientific directors need a unique set of skills that includes openness to different perspectives, the ability to conceptualize the big picture, and perhaps most importantly, a talent for uniting people around a common mission. These qualities are not always natural for scientists, so providing professional development opportunities to foster leadership in the transdisciplinary model is critical.

There are many examples of programs that already require transdisciplinary work: for example, grants provided by NSF's Innovations at the Nexus of Food, Energy and Water Systems (INFEWS) and the request for proposals outlined in the 2018 Sustainable Agricultural Systems competitive grants program administered through the U.S. Department of Agriculture's Agricultural and Food Research Initiative (USDA-AFRI). The NSF INFEWS and most recent USDA grants on Sustainable Agricultural Systems have relatively larger budgets that can support convergent team science. However, many of the standard grants requiring "transdisciplinary" approaches do not provide enough funding needed to support team science so incentives for transdisciplinary science are still lacking. For convergence to truly be productive, financial incentives are needed to encourage grant applicants to step outside of their comfort zones and to establish deep connections among subject matter experts from a variety of arenas.

Making the food system more sustainable and resilient can only be achieved through a better understanding of the system and its function. This is because the elements of the food system are highly interconnected, contain many complicated feedback loops, and are spatially and temporally heterogeneous. The biophysical elements of the food system are also tightly coupled with natural ecosystems, the built environment, and with human behaviors. Despite the fact that the food system will be inherently indeterminate at large scales, there are likely to be several critical nodes where a better understanding of the linkages would lead to significant improvement in overall system efficiency. The research community needs to come together to better define the food system, determine the barriers to successful development of integrated systems models of the food system, and identify approaches needed to overcome those barriers. These would need to be differentiated from other drivers of inefficiency, such as ineffective policies or undervaluation of natural resources. Such systems incorporate spatial and dynamic complexities and allow for synthesis of model outcomes and evaluation of trade-offs to alternative model and policy specifications (IOM and NRC, 2015). This greater understanding at the systems level can be translated into systems models to direct future research needs, or to determine which measurements are required to lower uncertainty in model parameters or formulations (e.g., feedback loops), or to identify policy interventions that could have the greatest overall system benefits. Given the complexity of the food system, it is desirable to conduct a decadal survey of modeling tools and databases to identify where progress has been made and where the greatest uncertainties remain.

A multidimensional systems approach will be essential for addressing policy and regulation. A systems approach can be used to consider the design and assess the effects of proposed agricultural policies and incentives to influence human behaviors that can support the desired systems-level sustainability outcomes (such as eliminating groundwater mining or mitigating nutrient loading to receiving streams). A systems approach will help to avoid unanticipated negative consequences to the environment or to humans from interventions in the food system. A systems approach will likely require a paradigm shift at all levels that encourages systems-based thinking and convergent approaches to problem solving.

Encouraging convergent research and improving our systems-level understanding of the food system will require changes in the way research is funded. NSF, USDA, the U.S. Department of Energy (DOE), and the U.S. Agency for International Development (USAID) are examples of four federal agencies with areas relevant to food and agriculture. There are already some examples of joint funding calls (e.g., NSF/USDA INFEWS) and requests for "convergence" research (e.g., 2018 USDA-AFRI grants on Sustainable Agricultural Systems and USDA/DOE/NASA-ROSES [National Aeronautics and Space Administration, Research Opportunities in Space and Earth Sciences] grants). Expanding these types of programs and deepening the connections between stakeholders and researchers from a wide range of disciplines—from physical and biological sciences to social and behavioral sciences to engineering—will promote the convergence required. Creating large, long-term (10 years or more), highly transdisciplinary research centers focused on a particular agriculture problem would also significantly promote convergence. Some examples of successful centers aimed at addressing a particular issue and constructed on integrated and collaborative approaches include USAID's Global Development Laboratory for Innovation, NSF's Long-Term Ecological Research Program, and NSF's Engineering Research Centers.

Most improvements in food system modeling will also require changes in the agri-food data cyberinfrastructure. Computational models depend on input data for development, validation, and application. The data requirements for integrated food systems models can be as intense as the food system under study is rich. All the inputs needed and outputs generated at each step involved in the process of growing, harvesting, processing, packaging, transporting, marketing, consuming, and disposing of food items within a social, political, economic, and environmental context are of interest. Experimental basic research provides essential data that otherwise do not exist. Making existing data findable, accessible, interoperable, and reuseable and encouraging open data (or other approaches to sharing proprietary data) is paramount to accelerating food system models.

Sensing Technologies

Breakthrough 2: The development and validation of precise, accurate, field-deployable sensors and biosensors will enable rapid detection and monitoring across various food and agricultural disciplines. Historically, sensors and sensing technology have been widely used in agriculture, food production, and distribution systems to provide point measurements for certain characteristics of interest (e.g., temperature). But the ability to monitor several characteristics at once is the key to understanding both what and how it is happening in the target system. Scientific and technological developments in materials science and nanotechnology are poised to enable the creation of novel nano- and biosensors to continuously monitor and detect levels and conditions of environmental stimuli and biotic and abiotic stresses. The next generation of sensors may also revolutionize the ability to detect disease prior to the onset of symptoms in plants and animals, to identify human pathogens before they enter the food distribution chain, and to monitor and make decisions in near real time.

Recommendation 2: Create initiatives to more effectively employ existing sensing technologies and to develop new sensing technologies across all areas of food and agriculture. These initiatives would lead to transdisciplinary research, development, and application across the food system.

The attributes of the sensor (e.g., shape, size, material, in situ or in planta, mobile, wired or wireless, and biodegradable) would depend on the purpose, application, duration, and location of the sensors. For example, in situ soil and crop sensors may provide continuous data feed and may alert the farmer when moisture content in soil and turgor pressure in plants fall below a critical level to initiate site-specific irrigation to a group of plants, eliminating the need to irrigate the entire field. Likewise, in planta sensors may quantify biochemical changes in plants caused by an insect pest or a pathogen, alerting and enabling the producer to plan and deploy immediate site-specific control strategies before infestation occurs or damage is visible. Biosensors for food products could indicate product adulteration or spoilage and could alert distributors and consumers to take necessary action.

Collaboration among scientists across various disciplines will be essential for developing the right set of characteristics for sensors. For instance, mobile sensors or wearable devices could transmit data to smartphones or other devices, and sensors would need to be developed that would be scale and target appropriate (nanoscale sensors for individual organisms [plants, animals] versus large-scale sensors for detecting soil characteristics). Sensing technologies would also need to be affordable to be implemented on a large scale and have the capability to be disposable even after single use at retail and consumer phases. The consideration of these various factors will require a convergence of disciplines (such as computer science, materials science, agronomy, food science, and animal sciences) to develop sensors that would be useful for field use and easily integrated into models and databases.

To facilitate progress, online or face-to-face platforms need to be developed that allow scientists from the various arenas (soil, plant, and animal) to become current on the latest technologies. Sharing across disciplines will spur new developments. Whether this is done through the institution of consortia, semiannual transdisciplinary meetings, or some other platform can be left to the researcher and policy community, again with stakeholder input.

One possibility might be to model sharing of sensor information and development on the European Viral Archive project, in which scientists from government and university laboratories from multiple countries worked together to create a virtual archive that would be available for the next emerging viral disease outbreak. After several years of collaboration and cooperation, an archive was available at the time of the Zika virus outbreak, and the archive allowed scientists from all over the world to make rapid progress on disease diagnosis and control.

Data Science and Agri-Food Informatics

Breakthrough 3: The application and integration of data science, software tools, and systems models will enable advanced analytics for

managing the food and agricultural system. Development of inexpensive field-deployable sensors, remote sensing capabilities, and omics techniques have resulted in the collection of enormous amounts of data, but the right tools have yet to be employed for using such data effectively. Data generated in research laboratories and in the field have been maintained in an unconnected manner, preventing the ability to generate insights from their integration. Advances and applications of data science and analytics have been highlighted as an important opportunity to elevate food and agricultural research and the application of knowledge. The ability to more quickly collect, analyze, store, share, and integrate highly heterogeneous datasets will create opportunities to vastly improve our understanding of the complex food system, and ultimately to the widespread use of near-real-time, data-driven management approaches.

Recommendation 3: Establish initiatives to nurture the emerging area of agri-food informatics and to facilitate the adoption and development of information technology, data science, and artificial intelligence in food and agricultural research. Data science and analytics are essential for addressing the most important challenges facing the food system. For example, data analytics that can rapidly link genotypes to phenotypes will help to provide the linkages required to select for desired traits in plants and animals, and will enable nutrigenomics research. Data-driven approaches and blockchain technologies that instantly transfer product data along the food supply chain can be employed to increase food quality and safety through real-time detection of pathogens. These same technologies can also be used to promote animal health, welfare, and productivity. Better analytics of disparate data sources will enable precision agriculture by using real-time data from distributed ground and remote sensing of soil moisture and nutrient levels, accurate weather predictions, plant and soil microbiome, and plant health data. Data collected at high spatial and temporal resolutions will enable scientists to better explore, model, and ultimately optimize the interactions between and functioning of complex systems.

Maximizing the knowledge and utility that can be gained from large research datasets requires strategic efforts to provide better data access, data harmonization, and data analytics in food and agricultural systems. The challenges of handling massive datasets that are highly heterogeneous across space and time need to be addressed. Data standards need to be established and the vast array of data need to be more findable, interoperable, and reuseable. There is a need to increase data processing speeds, develop methods to quickly assess data veracity, and provide support for the development and dissemination of agri-food informatics capabilities, including tools for modeling real-time applications in dynamically changing conditions.

Blockchain and artificial intelligence, including machine-learning algorithms, are promising technologies for the unique needs of the food and agricultural system that have yet to be fully developed. Development of advanced analytic approaches, such as machine-learning algorithms for automated rapid phenotyping, will require better platforms for studying how various components in the food system interact. Application of these approaches will require investment in infrastructure to house massive numbers of records and a means by which those records can be integrated and effectively used for decision-making purposes. A convergence of expertise from many disciplines will be needed to realize the potential of these opportunities.

Genomics and Precision Breeding

Breakthrough 4: The ability to carry out routine gene editing of agriculturally important organisms will allow for precise and rapid improvement of traits important for productivity and quality. Gene editing—aided by recent advances in omics (e.g., genomics, transcriptomics, proteomics, and metabolomics)—is poised to accelerate breeding to generate traits in plants, microbes, and animals that improve efficiency, resilience, and sustainability. Comparing hundreds of genotypes using omics technologies can speed the selection of alleles to enhance productivity, disease or drought resistance, nutritional value, and palatability. For instance, the tomato metabolome was effectively modified for enhanced taste, nutritional value, and disease resistance, and the swine genome was effectively targeted with the successful introduction of resistance to porcine reproductive and respiratory syndrome virus. This capability opens the door to domesticating new crops and soil microbes, developing disease-resistant livestock, controlling organisms' response to stress, and mining biodiversity for useful genes.

Recommendation 4: Establish initiatives to exploit the use of genomics and precision breeding to genetically improve traits of agriculturally important organisms. Genetic improvement programs in crops and animals are an essential component of agricultural sustainability. With the advent of gene-editing technologies, targeted genetic improvements can be applied to plant and animal improvement in a way that traditional methods of modification were unable to achieve. There are opportunities to accelerate genetic improvement by incorporating genomic information, advanced breeding technologies, and precision breeding methods into conventional breeding and selection programs. Encouraging the acceptance and adoption of some of these breakthrough technologies requires insight gained from social science and related education and communication efforts with producers and the public.

Gene editing could be used to both expand allelic variation introduced from wild relatives into crops and remove undesirable linked traits, thereby increasing the value of genetic variation available in breeding programs.

Similarly, incorporating essential micronutrients or other quality-related traits in crops through gene-editing tools offers an opportunity to increase food quality and shelf life, enhance nutrition, and decrease food loss and food waste. These technologies are similarly applicable to food animals, and possible targets of genetic improvements include enhanced fertility, removal of allergens, improved feed conversion, disease resistance, and animal welfare.

Genome sequencing and other omics technologies may enable diagnosis of unknown pathogens as well as pinpointing the disease-causing organism from a tissue sample, in a needle-in-a-haystack manner, all in real time. Further advances in this technology could enable rapid testing in the field and at low cost. Such a technology could markedly decrease the time to diagnose transboundary animal diseases and save millions (if not billions) of dollars over the current system of diagnostics.

More in-depth knowledge of omics and how they vary between organisms will be essential in devising robust strategies for detection. Especially in the realm of food safety, the ability to identify extremely small quantities of pathogens and the ability to determine strains could significantly enhance public health by identifying causes of foodborne outbreaks at the earliest possible moment. Detecting miniscule amounts of volatile materials—production of chemicals associated with spoilage—could lead to decreased food loss and waste and prompt preemptive removal to chilled conditions.

Microbiome

Breakthrough 5: Understand the relevance of the microbiome to agriculture, and harness this knowledge to improve crop production, transform feed efficiency, and increase resilience to stress and disease. Emerging accounts of research on the human microbiome provide tantalizing reports of the effect of resident microbes on the body's health. In comparison, a detailed understanding of the microbiomes in agriculture—animals, plants, and soil—is markedly more rudimentary, even as their functional and critical roles have been recognized for each at a fundamental level. A better understanding of molecular-level interactions between the soil, plant, and animal microbiomes could revolutionize agriculture by improving soil structure, increasing feed efficiency and nutrient availability, and boosting resilience to stress and disease. It is too early to draw conclusions about the relevance and potential applications of microbiomes across ecosystems of relevance to food and agriculture. However, with increasingly sophisticated tools to probe agricultural microbiomes, the next decade of research promises to bring increasing clarity to their role in agricultural productivity and resiliency.

Recommendation 5: Establish initiatives to increase the understanding of the animal, soil, and plant microbiomes and their broader applications across the food system. Transdisciplinary efforts focused on obtaining a better understanding of the various agriculturally relevant microbiomes and the complex interactions among them would create opportunities to modify and improve numerous aspects of the food and agricultural continuum. For example, understanding the microbiome in animals could help to more precisely tailor nutrient rations and increase feed efficiency. Knowing which microbes or consortia of organisms might be protective against infections could decrease disease incidence and/or severity and therefore lower losses. Research efforts are already under way to characterize the food microbiome in an effort to produce a reference database for microbes upon which rapid identification of human pathogens can be based. In plant sciences, research priorities are being established that focus on engineering various microbiomes to promote better disease control, drought resistance, and yield enhancement. Characterization of interactions between the soil and plant microbiomes (phytobiome) is critical. The soil microbiome is responsible for cycling of carbon, nitrogen, and many other key nutrients that are required for crop productivity, and carries out several other key ecosystem functions impacted in largely unknown ways by a changing climate. Enhanced understanding of the basic microbiome components and the roles they play in nutrient cycling is likely to be critical for ensuring continuing and sustainable crop production globally.

4. FURTHER CONSIDERATIONS

The science breakthroughs alone cannot transform food and agricultural research, as there are other factors that contribute to the success of food and agricultural research. Such factors include the research infrastructure, funding, and the scientific workforce. Other considerations include the social, economic, and political outcomes of various approaches.

4.1 Research Infrastructure Considerations

Conclusion 1: Investments are needed for tools, equipment, facilities, and human capital to conduct cutting-edge research in food and agriculture. Addressing agriculture's most vexing problems in a convergent manner will require investments in research infrastructure that facilitate convergence of disciplines on food and agricultural research. These could include physical infrastructure for experimentation as well as cyber infrastructure that enables sharing of ideas, data, models, and knowledge. Investments in our knowledge infrastructure are needed to develop a workforce capable of working in transdisciplinary teams and in a convergent manner. Mechanisms are also needed to facilitate building private–public partnerships and engaging the public in food and agricultural research. Some important infrastructure needs include

- Funded experimental facilities for crop, animal, agriculture, and food sciences where teams of scientists, engineers, companies, and other stakeholders can converge to test new methods and models and to engage the public at all steps of the process;
- Funding to encourage team science and to develop new educational programs that support convergence thinking and problem solving;
- Data platforms that are findable, accessible, interoperable, and reuseable and facilitate open access;
- Models that are open access and available to the whole research community; and
- Highly spatially and temporally resolved weather forecasting.

Conclusion 2: The Agricultural Experiment Station Network and the Cooperative Extension System deserve continued support because they are vital for basic and applied research and are needed to effectively translate research to achieve impactful results in the food and agricultural sectors. The agricultural sciences are grounded in the basic sciences but have an eye toward the applied; this has historically been facilitated by state agricultural experiment stations, as well as by extension and outreach efforts. Personnel and facilities with these functions allow scientists to translate laboratory-based findings into real-world products and processes that are most relevant, ultimately reaching key stakeholders and end users. Those stakeholders include industry, regulatory agencies, farmers and ranchers, and the general public. The recognition that scientists need to collaborate with stakeholders and translate basic research into useful and applicable results for the good of society is a fundamental value of the agricultural sciences. Recognizing and reinforcing that value through the provision of resources is essential for integrating agricultural scientific breakthroughs into the fabric of everyday life.

4.2 Funding Considerations

Conclusion 3: Current public and private funding for food and agricultural research is inadequate to address critical breakthrough areas over the next decade. There is a rapidly emerging need for food security and health to merit national priority and receive the funding needed to address the complex challenges in the next decade. If a robust food system is critical for securing the nation's health and well-being, then funding in both the public and private sectors ought to reflect this as a priority.

In the past century, public funding for food and agricultural research has been essential for enabling talented U.S. scientists to conduct basic scientific research and provide innovative solutions for improving food and agriculture. Agricultural R&D by the public sector provides benefits that accrue to both farmers as well as consumers and is estimated to have a median rate of return of 40 percent (Clancy et al., 2016). Waning U.S. public investments will slow innovation and growth and would jeopardize the ability of the United States to remain competitive in a global economy, potentially undermining U.S. food and nutrition security (NRC, 2014b). As previously mentioned, since 2009, China has surpassed the United States as the top global performer of agricultural R&D (Clancy et al., 2016). In fiscal year (FY) 2017, the National Institutes of Health (NIH) allocated $18.2 billion for competitive research grants compared to USDA, which was appropriated only $325 million for competitive research grants (less than 2 percent of the NIH's amount), a budget that was less than half of the congressionally authorized amount (HHS, 2017; USDA, 2017). The current level of federal funding for food and agricultural research has thus been inadequate. Breakthrough science needed to assist the food and agricultural enterprise to thrive in the future will require a significant investment. More will be required to sustain the level of coordination and collaboration needed to address the increasingly integrative, expansive, and visionary research required to ensure future security and competitiveness.

The current political climate suggests that it may be difficult to increase public funding for food and agricultural research to the levels needed. Although private R&D is not a substitute for public R&D funding, private foundations and industry can provide some research funding that is complementary to public funding in the U.S. agricultural innovation system. Innovative business models can be more widely employed for engaging researchers. For example, venture capital funding for start-up companies, which are well known in the tech industry, are providing record sources of investment in food and agricultural research (The Context Network, 2017; Cosgrove, 2018; Rausser et al., 2018). There are new institutions and mechanisms of financing research and of implementing innovations induced by research that offer the potential to expand funding. For instance, the Foundation for Food and Agriculture Research was established by Congress in the Agricultural Act of 2014 (known as the 2014 Farm Bill) to increase investment in cutting-edge research by leveraging public funds with matching nonfederal dollars (including industry and nonprofit organizations) and thus expand the total funding available to support research. Other examples of new institutions include the Agricultural Technology Innovation Partnership Foundation formed to develop public–private collaborations around USDA research discoveries technology transfer, including linkage to venture capital and business expertise. However, these sources alone are insufficient to achieve the goals laid out in this report. In order for the U.S. agricultural enterprise to capitalize on the integrative, expansive, and visionary tools of research now being actively pursued by many other industries (e.g., sensing technologies, wireless communication, and machine learning), a commitment to a major investment is needed now to ensure their relevant application to food and agriculture.

4.3 Education and Scientific Workforce

Conclusion 4: Efforts to renew interest in food and agriculture will need to be made to engage nonagricultural professionals and to excite the next generation of students. Vast opportunities are available for non-traditional agriculture professionals to be involved in food and agriculture. However, there may be barriers to their involvement, such as misperceptions about the sophistication of agricultural technology and the lack of sustained funding for building transdisciplinary agricultural research teams that include non-agricultural professionals and scientists from other disciplines to work in food and agricultural sciences.

A robust workforce for food and agricultural research will require talented individuals who are proficient in the challenges facing the food system along with an understanding of the opportunities to think outside the box for innovative approaches. Recruiting talented individuals into food and agricultural research will require a demonstration and shift in perception that food and agriculture can be innovative. There are efforts under way to merge traditional agricultural disciplines with other areas of expertise to bring about convergent approaches, such as the new undergraduate program in computer science and crop sciences at the University of Illinois at Urbana-Champaign beginning in fall 2018 (UIUC, 2017).

4.4 Socioeconomic Contributions and Other Considerations

Conclusion 5: A better understanding of linkages between biophysical sciences and socioeconomic sciences is needed to support more effective policy design, producer adoption, and consumer acceptance of innovation in the food and agricultural sectors. The successful application of scientific innovation in the field depends on the willingness and ability of stakeholders to successfully apply and use new products and processes; it also depends on whether they view high-tech, site-specific approaches as economically or ecologically beneficial. There is a critical need to better understand the best means and methods for effective technology development and integration in production processes, with input from both the public and private sectors. Better understanding of the political economy, behavioral and choice processes related both to adoption and use of the technological innovation, and acceptance and perception of new products will be required to support the effective design of policies and application of the research innovation (Herring and Paarlberg, 2016; Clancy and Moschini, 2018). For example, digital information from remote sensing devices may be used in new decision support systems to assist agricultural workers in making choices about field practices or animal handling. However, workers will need sufficient training and motivation to respond to expected and unexpected outcomes and uncertainty (e.g., animal response to treatment or extreme weather events). Lessons from behavioral sciences (e.g., "nudges" or defaults) may help support behavioral change and training requirements. An example of "nudges" or defaults is in the reduced costs of adoption and use of water conservation practices (Ferraro et al., 2017). Attention to the political economy and socioeconomic context highlights the challenge of distributional aspects as well. Small-scale operators may be limited in their ability to take advantage of newer technologies because of cost considerations or capital requirements. Some consumers may be limited in food choices by lack of access to product variety or innovative data applications in their local food outlets and environment. However, to allow more rapid diffusion of the technological advances and offer an important way to address issues of potential scale bias in some of the new technologies, innovations are needed in scaling the available technologies and in the market for services that provide user friendly access to data services for small-scale producers or consumers through devices and apps.

The successful implementation of scientific advances also requires other important considerations to be taken into account. Policies on land or input use, environmental impact, animal welfare, and food-handling practices can have significant near- and long-term impacts on agricultural and food sustainability. Some policy or technology changes may have unintended consequences in the system and require closer examination of system interactions, including human behaviors related to adoption and use of new inputs, products, and processes. Insights from behavioral sciences can help inform policy designs and reduce the costs of change, inform technological adoption in the field (e.g., design of conservation or tillage applications, or provision of product information to consumers), and address issues of product acceptance and consumer trust in the food system (Lusk and McCluskey, 2018).

5. CLOSING REMARKS

At this pivotal time in history with an expanding global population requiring more from an increasingly fragile natural resource base, science breakthroughs are needed now more than ever. As the world's greatest agricultural producer, the United States bears the tremendous responsibility of implementing scientific advances to support our nation's well-being and security, and perhaps even global stability. Promising research opportunities include integrating agriculture and food systems to sustainably meet human and animal health needs, providing yield stability and economic value under variable and uncertain environmental pressures, and reducing inputs and negative environmental impacts. Realizing this vision requires a holistic systems approach that combines scientific discovery, technological innovation, and incentives to revolutionize our food and agricultural systems to ensure greater food security and human and environmental health. Implementing this vision requires a bolder approach to research that integrates scientific insight from various disciplines to bring promising breakthroughs to fruition and to ensure U.S. competitiveness and food security. The food system of tomorrow will depend on how well we are able to prepare for resiliency today and how well we are able to build our capacity for the future. The U.S. scientific enterprise is willing to rise to address such challenges; the tools and resources identified in this report can ensure its success.

REFERENCES

Buzby, J. C., F. W. Hodan, and J. Hyman. 2014. *The Estimated Amount, Value, and Calories of Postharvest Food Losses at the Retail and Consumer Levels in the United States.* Economic Information Bulletin No. EIB-121. Washington, DC: USDA Economic Research Service. Available at https://www.ers.usda.gov/publications/pub-details/?pubid=43836 (accessed June 11, 2018).

Clancy, M., and G. Moschini. 2018. Mandates and the incentive for environmental innovation. *American Journal of Agricultural Economics* 100(1):198-219.

Clancy, M., K. Fuglie, and P. Heisey. 2016. *U.S. Agricultural R&D in an Era of Falling Public Funding.* U.S. Department of Agriculture, Economic Research Service. Available at https://www.ers.usda.gov/amber-waves/2016/november/us-agricultural-rd-in-an-era-of-falling-public-funding (accessed May 8, 2018).

The Context Network. 2017. Private Equity and Venture Capital Investment in Ag Tech Grow Despite Headwinds. Available at https://contextnet.com/private-equity-and-venture-capital-investment-in-ag-tech-grow-despite-headwinds (accessed June 11, 2018).

Cosgrove, E. 2018. AgriFood Tech Investment Surges to $10.1 Billion Bringing a New Normal. Available at https://agfundernews.com/agrifood-investment-surges-10bn-new-normal.html (accessed June 11, 2018).

Crowley, K., and B. W. Head. 2017. The enduring challenge of "wicked problems": Revisiting Rittel and Webber. *Policy Sciences* 50:539-547.

Ferraro, P., K. D. Messer, and S. Wu. 2017. Applying behavioral insights to improve water security. *Choices* (4). Available at http://www.choicesmagazine.org/choices-magazine/theme-articles/inducing-water-conservation-in-agriculture-institutional-and-behavioral-drivers/applying-behavioral-insights-to-improve-water-security (accessed June 11, 2018).

Herring, R., and R. Paarlberg. 2016. The political economy of biotechnology. *Annual Review of Resource Economics* 8:397-416.

HHS (U.S. Department of Health and Human Services). 2017. HHS FY 2017 Budget in Brief—NIH. Available at https://www.hhs.gov/about/budget/fy2017/budget-in-brief/nih/index.html (accessed May 9, 2018).

IOM and NRC (Institute of Medicine and National Research Council). 2015. *A Framework for Assessing Effects of the Food System.* Washington, DC: The National Academies Press.

Khanna, M., and D. Zilberman. 2017. (Theme overview) Inducing water conservation in agriculture: institutional and behavioral drivers. *Choices* (4). Available at http://www.choicesmagazine.org/choices-magazine/theme-articles/inducing-water-conservation-in-agriculture-institutional-and-behavioral-drivers/theme-overview-inducing-water-conservation-in-agriculture-institutional-and-behavioral-drivers (accessed June 11, 2018).

Lusk, J. L., and J. McCluskey. 2018. Understanding the impacts of food consumer choice and food policy outcomes. *Applied Economic Perspectives and Policy* 40(1):5-21.

NRC (National Research Council). 2014a. *Convergence: Facilitating Transdisciplinary Integration of Life Sciences, Physical Sciences, Engineering, and Beyond*. Washington, DC: The National Academies Press.

NRC. 2014b. *Spurring Innovation in Food and Agriculture: A Review of the USDA Agriculture and Food Research Initiative Program*. Washington, DC: The National Academies Press.

NSF (National Science Foundation). 2018. Convergence Research at NSF. Available at https://www.nsf.gov/od/oia/convergence/index.jsp (accessed April 10, 2018).

Perry, C., P. Steduto, and F. Karajeh. 2017. Does Improved Irrigation Technology Save Water? A Review of the Evidence. Cairo, Egypt: UN Food and Agriculture Organization. http://www.fao.org/3/I7090EN/i7090en.pdf.

Rausser, G., B. Gordon, and J. Davis. 2018. Recent Developments in the California Food and Agricultural Technology Landscape. Available at https://giannini.ucop.edu/publications/are-update/issues/2018/21/4/recent-developments-in-the-california-food-and-agr (accessed June 12, 2018).

Rittel, H. W. J., and M. M. Webber. 1973. Dilemmas in a general theory of planning. *Policy Sciences* 4(2):155-169.

Steward, D. R., P. J. Bruss, X. Yang, S. A. Staggenborg, S. M. Welch, and M. D. Apley. 2013. Tapping unsustainable groundwater stores for agricultural production in the High Plains Aquifer of Kansas, projections to 2110. *Proceedings of the National Academy of Sciences of the United States of America* 110(37):E3477-E3486.

Taylor, R., and D. Zilberman. 2017. Diffusion of drip irrigation: The case of California. *Applied Economic Perspectives and Policy* 39(1):16-40.

UIUC (University of Illinois at Urbana-Champaign). 2017. New Degree in Computer Science Plus Crop Sciences Melds the Worlds of Agriculture and Data Technology. Available at https://news.aces.illinois.edu/news/new-degree-computer-science-plus-crop-sciences-melds-worlds-agriculture-and-data-technology (accessed June 11, 2018).

USDA (U.S. Department of Agriculture). 2017. FY 2017 Budget Summary. Available at https://www.obpa.usda.gov/budsum/fy17budsum.pdf (accessed May 10, 2018).

Appendix A

Biographical Sketches of Committee Members

CO-CHAIRS

Dr. John D. Floros currently serves as President of New Mexico State University (NMSU). Prior to his appointment to NMSU in 2018, Dr. Floros was a professor of food science and engineering, Dean of the College of Agriculture, and Director of K-State Research and Extension at Kansas State University. He was also Head of the Department of Food Science at The Pennsylvania State University, worked in the food industry, and served on several boards, including the Food and Drug Administration's Science Board. He is a Fellow and Past President of the Institute of Food Technologists (IFT), and a Fellow of the Food Systems Leadership Institute. He has expertise in food processing, engineering, and packaging systems, and a broad understanding of food safety, microbiology, biochemistry, and material science. As dean, he gained a broader perspective of our complex food system, including knowledge of animal and plant agriculture, water and natural resources, and energy and environmental issues, as well as social and behavioral concerns. Similarly, he understands the continuum of agriculture–food–nutrition–health, and he has been extensively involved in international agriculture efforts and improving the sustainability of the global food system. He has published more than 220 refereed articles, research abstracts, book chapters, and other publications, presented more than 400 scientific lectures, was invited to more than 300 lectures, and received numerous awards and professional honors throughout his career. Dr. Floros received his B.S./M.S. in food science and technology from the

Agricultural University of Athens, Greece, and his Ph.D. in food and science technology from the University of Georgia, Athens.

Dr. Susan R. Wessler (NAS) is currently the Neil and Rochelle Campbell Presidential Chair for Innovation in Science Education and Distinguished Professor of Genetics at the University of California, Riverside. She is a member of the National Academy of Sciences and in 2011 she was elected as its Home Secretary, the first woman to hold this position in its 150-year history. She is also a fellow of the American Association for the Advancement of Science, the American Academy of Arts and Sciences, the American Philosophical Society, and a Foreign Member of the Royal Society. Dr. Wessler is a plant molecular geneticist who studies the role of transposable elements in generating genetic diversity. Her laboratory has shown that transposable elements are an important mutagenic force fueling plant gene and genome evolution. She discovered a new type of transposon, called MITES, and unraveled key features of gene regulation through her comparative studies of rice and maize. Dr. Wessler has contributed extensively to educational initiatives. As a Howard Hughes Medical Institute Professor, she adapted her research program for the classroom by developing the Dynamic Genome Courses where incoming freshman can experience the excitement of scientific discovery. Dr. Wessler is the recipient of several awards, including the inaugural Distinguished Scientist Award from the Southeastern Universities Research Association (SURA), the Stephen Hales Prize from the American Society of Plant Biologists, the Excellence in Science Award from the Federation of American Societies for Experimental Biology, and the McClintock Prize for Plant Genetics and Genome Studies from the Maize Genetics Community. She earned her B.A. in biology from the State University of New York at Stony Brook in 1974 and her Ph.D. in biochemistry from Cornell University in 1980.

COMMITTEE MEMBERS

Dr. David B. Allison (NAM) is Dean and Provost Professor at the Indiana University–Bloomington School of Public Health. Dr. Allison received his Ph.D. from Hofstra University in 1990. He then completed a postdoctoral fellowship at the Johns Hopkins University School of Medicine and a second postdoctoral fellowship at the National Institutes of Health (NIH)-funded New York Obesity Research Center at St. Luke's/Roosevelt Hospital Center. He was a research scientist at the New York Obesity Research Center and associate professor of medical psychology at Columbia University College of Physicians and Surgeons until 2001. Prior to becoming Dean and Provost Professor at the Indiana University–Bloomington School of Public Health in 2017, he was Distinguished Professor, Quetelet Endowed Professor, and Director of the NIH-funded Nutrition Obesity Research Center (NORC) at the University of Alabama at Birmingham. He has authored more than 500 scientific publications and received many awards. In 2012 he was elected to the National Academy of Medicine. He has served on the Scientific Advisory Board for the Nutrition Science Initiative (NuSI) and currently serves on the board-appointed Committee on Science and Technology Engagement with the Public (CoSTEP) for the American Association for the Advancement of Science, 2014-2020. He serves or has served on many editorial boards and currently serves as associate editor or statistical editor for *Obesity*; *International Journal of Obesity*; *Nutrition Today*; *Obesity Reviews*; *Public Library of Science (PLoS) Genetics*; *Surgery for Obesity and Related Diseases (SOARD)*; and *American Journal of Clinical Nutrition*. Dr. Allison is also proud to be the founding Field Chief Editor of *Frontiers in Genetics*. Dr. Allison's research interests include obesity and nutrition, quantitative genetics, clinical trials, statistical and research methodology, and research rigor and integrity.

Dr. Corrie C. Brown is a Josiah Meigs and University Distinguished Professor of Anatomic Pathology in the College of Veterinary Medicine at the University of Georgia. Dr. Brown received her Ph.D. in veterinary pathology with a specialization in infectious diseases from the University of California, Davis, and her D.V.M. from the University of Guelph. She is a Diplomate of the American College of Veterinary Pathologists. Her research interests focus on pathogenesis of infectious disease in food-producing animals, especially the transboundary animal diseases. She is active in the fields of emerging diseases and international veterinary medicine. She has published or presented more than 300 scientific papers. Dr. Brown has served on many industrial and federal panels, and has been a technical consultant to several federal agencies and intergovernmental organizations on issues involving infectious diseases and animal health infrastructure. Dr. Brown has received numerous awards for her teaching and service at the college, university, and national levels.

Dr. Lisa M. Goddard is the Director of the International Research Institute (IRI) for Climate and Society and an adjunct associate professor within the Department of Earth and Environmental Sciences at Columbia University. She has been involved in El Niño and climate forecasting research and operations since the mid-1990s. She has extensive experience in forecasting methodology and has published papers on El Niño, seasonal climate forecasting and verification, and probabilistic climate change projections. Currently leading the IRI's effort on near-term climate change, Dr. Goddard oversees research and product development aimed at providing climate information at the 10- to 20-year horizon and how that low frequency variability and change interacts with the probabilistic risks and benefits of seasonal-to-interannual variability. Most of Dr. Goddard's research focuses on diagnosing and extracting meaningful information from climate models and available observations. She also developed and oversees a new national postdoctoral program, the Post-docs Applying Climate Expertise Program (PACE), which explicitly links recent climate doctorate graduates with decision-making institutions. Dr. Goddard holds a Ph.D. in atmospheric and oceanic sciences from Princeton University and a B.A. in physics from the University of California, Berkeley.

Dr. Mary Lou Guerinot (NAS) is the Ronald and Deborah Harris Professor in the Department of Biological Sciences at Dartmouth College. In 2016, she was elected to the National Academy of Sciences. Dr. Guerinot pioneered research on metal metabolism in plants through key discoveries of genes involved in major transport processes for minerals such as iron and zinc. Her research is critically important for both agriculture and human nutrition since iron and zinc deficiencies affect billions of humans that rely on crop-based diets. She received her B.S. in biology from Cornell University, Ph.D. in biology from Dalhousie University, and completed two postdoctoral fellowships at the University of Maryland, College Park, and the Department of Energy-Plant Research Laboratory, Michigan State University. She currently serves as the Chair of the Scientific Advisory Board for the Boyce Thompson Institute and is on the Board of Directors for the Genetics Society of America. She is a recipient of the Dartmouth Graduate Mentoring Award and the Dean of Faculty Award for Outstanding Mentoring and Advising, the Dennis R. Hoagland Award, and Stephen Hales Prize from the American Society for Plant Biologists.

Dr. Janet K. Jansson is the Chief Scientist for Biology in the Earth and Biological Sciences Directorate and a Laboratory Fellow at the Pacific Northwest National Laboratory (PNNL). Dr. Jansson's research interests are in the application of molecular "omics" tools to gain an understanding of the function of microbial communities in complex environments, ranging from soil to the human gut. She is currently coordinating two large research initiatives at PNNL; one is focused on microbiomes in transition, "MinT," and the other is a U.S. Department of Energy–funded project on the soil microbiome. From 2007 to 2014, she was a Senior Staff Scientist and at the Lawrence Berkeley National Laboratory and an adjunct professor at the University of California, Berkeley. Prior to that, she spent 20 years in Sweden with her last position as Professor (Chair) of Environmental Microbiology at the Swedish University of Agricultural Sciences (SLU) and Vice Dean of the Natural Science Faculty. Dr. Jansson recently completed a term as the President of the International Society for Microbiology. She is a Fellow of the American Academy of Microbiology and of the Washington State Academy of Science. Dr. Jansson received her Ph.D. (1988) in microbial ecology from Michigan State University.

Dr. Lee-Ann Jaykus is a William Neal Reynolds Distinguished Professor in the Department of Food, Bioprocessing, and Nutrition Sciences at North Carolina State University. Her current research efforts are varied and focus on (1) food virology, (2) development of molecular methods for foodborne pathogen detection, (3) application of quantitative risk assessment in food safety, and (4) understanding the ecology of pathogens in foods. She is currently serving as the scientific director of the U.S. Department of Agriculture's National Institute of Food and Agriculture Food Virology Collaborative. Also called NoroCORE, the Collaborative is a large consortium of scientists and stakeholders working collectively to reduce the burden of foodborne illness associated with viruses. Her professional activities have included membership on the National Advisory Committee on Microbiological Criteria for Foods; participation in several National Academies of Sciences, Engineering, and Medicine consensus studies and as a member of the Food and Nutrition Board and the Food Forum; and on the executive board of the International Association for Food Protection, for which she served as president in 2010-2011. She has taught food microbiology/safety at the undergraduate and graduate levels, has mentored more than 40 graduate students and 15 postdoctoral research associates, and authored or co-authored more than 170 publications. Dr. Jaykus received a B.S. degree in food science and an M.S. in animal science (food microbiology) from Purdue University. Her Ph.D. is from the University of North Carolina at Chapel Hill School of Public Health.

Dr. Helen H. Jensen is a professor of economics and leads a research group focused on food and nutrition programs in the Center for Agricultural and Rural Development at Iowa State University, an internationally recognized research center that addresses issues of the food, agricultural, and natural resource sectors. Her current research focuses on the design of food and nutrition programs and policies, assessment of nutritional enhancement of foods, food demand and markets, linkages between agricultural policies and nutrition, and food safety regulations. She has led projects that analyze food demand; involve dietary, nutritional, and health assessment; and implement food consumption surveys in the United States as well as in several developing countries. Dr. Jensen was elected a Fellow of the Agricultural and Applied Economics Association (AAEA) in 2012 and has served on the Executive Board of AAEA and the Council on Food, Agriculture, and Resource Economics (C-FARE). She has served on several committees of the National Academies of Sciences, Engineering, and Medicine, including the Committee to Review Women, Infants, and Children Food Packages. She is an active member of the Food Forum. Dr. Jensen received her B.A. in economics from Carleton College, M.S. in applied economics from the University of Minnesota, and Ph.D. in agricultural economics from the University of Wisconsin–Madison.

Dr. Rajiv Khosla is a Robert Gardner Professor of Precision Agriculture at Colorado State University. His main research focus has been on management of in-field soil and crop variability using geospatial technologies for precision management of crop inputs. He has generated many discoveries in precision agriculture; most widely recognized among them is the innovative technique of quantifying variability of spatially diverse soils using satellite based remote sensing to create management zones. Most recently, he was recognized with the Werner L. Nelson Award for Diagnosis of Yield-Limiting Factors by the American Society of Agronomy. Previously, in 2015, he was recognized as the "Precision Ag Educator of the Year," a national honor bestowed by the agricultural industry, and in 2012 he was named the Jefferson Science Fellow by the National Academy of Sciences. Previously, he has served two 2-year terms on the National Aeronautics and Space Administration's U.S. Presidential Advisory Board on Positioning, Navigation and Timing. He is a Fellow of the American Society of Agronomy; the Soil Science Society of America; the Soil and Water Conservation Society; and an Honorary Life Fellow of the International Society of Precision Agriculture. He is the Founder and Founding President of the International Society of Precision Agriculture. Dr. Khosla received his B.S. in agricultural sciences at the University of Allahabad, India, M.S. in soil physics from Virginia Tech, and Ph.D. in soil fertility and crop management from Virginia Tech.

Dr. Robin Lougee is the IBM Research Lead for Consumer Products & Agriculture and a member of the IBM Industry Academy. She is the Chair of the Steering Committee of the Consultative Group on International Agricultural Research Platform for Big Data in Agriculture and the founding chair of the Syngenta Crop Challenge in Analytics Prize awarded by the Institute for Operations Research and the Management Sciences (INFORMS) Analytics Society. Dr. Lougee serves on the Advisory Council for the Food Science Department at Cornell University and the Advisory Committee for the World Agri-Tech Innovation Summit. She is an industrial research scientist with a strong track record of delivering innovation to IBM and its customers. Dr. Lougee pioneered the creation of Computational Infrastructure for Operations Research, an open-source foundry for computational operations research, and led its growth as an independent nonprofit that has served the scientific and business community for more than 15 years. She was elected to the Board of INFORMS, the largest society in the world for professionals in the field of operations research, management science and analytics, Chair of the INFORMS Computing Society, and President of the Fora of Women in Operations Research and the Management Sciences. Dr. Lougee is a past Associate Editor of *Surveys in Operations Research*. She earned a Ph.D. in mathematical sciences from Clemson University in 1993.

Dr. Gregory V. Lowry is the Walter J. Blenko, Sr. Professor of Civil and Environmental Engineering at Carnegie Mellon University. He is the Deputy Director of the National Science Foundation Environmental Protection Agency Center for Environmental Implications of Nanotechnology. He was a founding Associate Editor of *Environmental Science: Nano,* and currently serves on the editorial boards of *Environmental Science: Nano* and *Nature: Scientific Data*. His research aims to safely harness the unique properties of engineered nanomaterials for making water treatment and crop agriculture more sustainable. Recent work aims at understanding how a nanomaterial's properties and environmental conditions influence its fate in soils, nanomaterial–plant interactions, nutrient uptake efficiency, and disease management. He has authored more than 140 peer-reviewed journal articles. He has served as principal investigator (PI) or co-PI on grants from the National Science Foundation, U.S. Department of Defense, U.S. Department of Energy, and U.S. Environmental Protection Agency (EPA), as well as from industry. He served on the EPA Science Advisory Board (Environmental Engineering committee) and currently serves on the board of directors of the Association of Environmental Engineering and Science Professors. He served on the National Research Council Committee to Develop a Research Strategy for Environmental, Health, and Safety Aspects of Engineered Nanomaterials. Dr. Lowry holds a B.S. in chemical engineering from the University of California, Davis, an M.S. from the University of Wisconsin–Madison, and a Ph.D. in civil and environmental engineering from Stanford University.

Dr. Alison L. Van Eenennaam is a Cooperative Extension Specialist in the field of animal genomics and biotechnology in the Department of Animal Science at University of California, Davis (UC Davis), where she has been on faculty for 15 years. Her publicly funded research and outreach program focuses on the use of animal genomics and biotechnology in livestock breeding and production systems. Her current research projects include the development of genomic approaches to select for cattle that are less susceptible to disease, the development of genome editing approaches for livestock, and applied uses of DNA-based information on commercial beef cattle operations. She has given more than 550 invited presentations to audiences globally, and uses a variety of media to inform general public audiences about science and technology. Dr. Van Eenennaam was the recipient of the 2014 Council for Agricultural Science and Technology Borlaug Communication Award, and in 2017 was elected as a Fellow of the American Association for the Advancement of Science. She received a Bachelor of Agricultural Science from the University of Melbourne in Australia and both an M.S. in animal science and a Ph.D. in genetics from UC Davis.

Appendix B

Open Session Meeting Agendas

MEETING 1 AGENDA
June 14, 2017
National Academy of Sciences Building, Room 125

WEDNESDAY, June 14

OPEN SESSION

1:00–1:15 p.m.	**Welcome and Introductions** *Dr. John D. Floros and Dr. Susan R. Wessler, Committee Co-Chairs*
1:15–1:30 p.m.	**National Academies of Sciences, Engineering, and Medicine's Study Process and Committee's Statement of Task** *Mrs. Peggy T. Yih, Study Director*
1:30–1:40 p.m.	**Origins of Study Request** *Dr. Robert Easter, University of Illinois (Emeritus)*
1:40–2:10 p.m.	**Charge to the Committee from the Sponsors** *Dr. Robert Easter, University of Illinois (Emeritus)* *Dr. Sally Rockey, Foundation for Food and Agriculture Research*

2:10–2:50 p.m.	**Potential for the Study to Advance Food and Agricultural Research: U.S. Department of Agriculture Panel** *Dr. Mary Bohman, USDA Economic Research Service* *Dr. Meryl Broussard, USDA National Institute of Food and Agriculture* *Dr. Steve Kappes, USDA Agricultural Research Service* *Dr. Joe Parsons, USDA National Institute of Food and Agriculture* *Dr. Dionne Toombs, USDA Office of the Chief Scientist*
2:50–3:00 p.m.	**Break**
3:00–3:30 p.m.	**Science Frontiers of Interest** *Dr. Todd Anderson, Department of Energy* *Dr. Jane Silverthorne, National Science Foundation*
3:30–3:50 p.m.	**Decadal Vision for Plant Biology** *Mr. Tyrone Spady, American Society of Plant Biologists*
3:50–4:10 p.m.	**Challenge of Change—Recently Released Report** *Dr. Samantha Alvis, Association of Public & Land-Grant Universities*
4:10–4:30 p.m.	**Agricultural and Applied Economic Priorities and Solutions** *Ms. Caron Gala, Council on Food, Agricultural & Resource Economics*
4:30–4:50 p.m.	**Nutrition, Health, and Agriculture** *Dr. Daniel J. Raiten, National Institutes of Health*
4:50–5:10 p.m.	**Public Comments**
5:10–5:15 p.m.	**Co-Chairs' Closing Remarks** *Dr. John D. Floros and Dr. Susan R. Wessler, Committee Co-Chairs*
5:15 p.m.	**Adjourn Meeting for Day 1**

TOWN HALL AGENDA
August 8, 2017
National Academy of Sciences Building, Fred Kavli Auditorium

TUESDAY, August 8

OPEN SESSION

7:30–8:25 a.m. **Working Breakfast for Committee and Panelists (Members' Room)**

8:30–8:45 a.m. **Co-Chairs' Welcome Remarks**
Dr. John D. Floros and Dr. Susan R. Wessler, Committee Co-Chairs

8:45–10:30 a.m. **Panel 1: Food Production**
Lead speaker will provide a 5-minute overview, and will then provide 10-minute remarks about the greatest challenges and opportunities in their area of expertise. Each panelist will provide 5- to 10-minute formal remarks about the greatest challenges and opportunities in their area of expertise. After panelists have provided formal remarks, there will be a moderated discussion with panelists, committee, and members of the audience (in-person and online).

Lead speaker: Overview of Food Production and Animal Science Priorities
Dr. Ronnie Green, University of Nebraska (via videoconference)

Improving Photosynthetic Efficiency for Improved Yield
Dr. Donald Ort, U.S. Department of Agriculture Agricultural Research Service and University of Illinois

Postharvest Reduction of Food Waste: A Magic Bullet?
Dr. Daryl Lund, University of Wisconsin–Madison

	Computing Reimagined *Dr. Tim Dalton, IBM Watson*
	Science of Communication *Dr. Dietram Scheufele, University of Wisconsin–Madison*
	Moderated Discussion (with panelists, committee, and members of the audience) *Co-Moderators: Dr. Robin Lougee and Dr. Alison Van Eenennaam, Committee Members*
10:30–10:40 a.m.	**Key Take-Away Messages from Panel Discussions** *Dr. John D. Floros, Committee Co-Chair*
10:40–10:55 a.m.	**Break** *Coffee break in Great Hall*
10:55–11:15 a.m.	**The Promise of Plant Probiotics: A Potential Ag Revolution** *Dr. Jeff Dangl, University of North Carolina at Chapel Hill (via videoconference)*
11:15–11:35 a.m.	**Diets, Environmental Sustainability, and Human Health** *Dr. David Tilman, University of Minnesota (via videoconference)*
11:35 a.m.–12:00 p.m.	**Public Comments**
12:00–1:00 p.m.	**Lunch** *Working lunch for committee, speakers, and panelists in Members' Room*
1:00–2:45 p.m.	**Panel 2: Sustainability and Efficiency** *Lead speaker will provide a 5-minute overview and will then provide 10-minute remarks about the greatest challenges and opportunities in their area of expertise. Each panelist will provide 5- to 10-minute formal remarks about the greatest challenges and opportunities in their area of expertise. After panelists have provided formal*

remarks, there will be a moderated discussion with panelists, committee, and members of the audience (in-person and online).

Lead speaker: Convergence of Innovations for Sustainable Outcomes
Dr. James Jones, University of Florida

Soils
Dr. Rattan Lal, The Ohio State University

The Great Nitrogen Imbalance
Dr. Phil Robertson, Michigan State University

Designing for Sustainable and Resilient Human Environmental Agricultural Systems
Dr. Meagan Mauter, Carnegie Mellon University

Moderated Discussion (with panelists, committee, and members of the audience)
Co-Moderators: Dr. Greg Lowry and Dr. Mary Lou Guerinot, Committee Members

2:45–2:55 p.m. **Key Take-Away Messages from Panel Discussions**
Dr. Susan R. Wessler, Committee Co-Chair

2:55–3:10 p.m. **Break**
Coffee break in Great Hall

3:10–4:55 p.m. **Panel 3: Human Health**
Lead speaker will provide a 5-minute overview, and will then provide 10-minute remarks about the greatest challenges and opportunities in their area of expertise. Each panelist will provide 5- to 10-minute formal remarks about the greatest challenges and opportunities in their area of expertise. After panelists have provided formal remarks, there will be a moderated discussion with panelists, committee, and members of the audience (in-person and online).

	Lead speaker: Nutrition and Agricultural Production: Human Health as the Ultimate Mission *Dr. Pamela Starke-Reed, USDA-ARS*
	Nutrition, Food, and Health *Dr. Barbara Schneeman, University of California, Davis*
	Bridging Two Worlds with Technology and Networks *Dr. Mary Torrence, Food and Drug Administration*
	Importance of Understanding Behavioral Responses to Food and Health Policies *Dr. Jayson Lusk, Purdue University*
	Moderated Discussion (with panelists, committee, and members of the audience) *Moderator: Dr. Helen Jensen, Committee Member*
4:55–5:05 p.m.	**Key Take-Away Messages from Panel Discussions** *Dr. John D. Floros, Committee Co-Chair*
5:05–5:25 p.m.	**Public Comments** Please sign up in person at the registration table
5:25–5:30 p.m.	**Closing Remarks** *Dr. John D. Floros and Dr. Susan R. Wessler, Committee Co-Chairs*
5:30 p.m.	**Adjourn Meeting**

JAMBOREE WORKSHOP AGENDA
October 2-4, 2017
Beckman Center, Irvine, CA

MONDAY, October 2

8:30 a.m.	Auditorium	Welcome and Introduction, Committee Co-Chairs
8:45 a.m.		Jamboree Participant Warm-Up: Lightning Round Questions

10:15 a.m.		Summary of Lightning Round and Charge to the Group, Committee Co-Chairs
10:30 a.m.	Atrium	Break
10:45 a.m.	Auditorium	Keynote: A Vision for the Future of Agricultural Research *Cathie Woteki, former Chief Scientist and Undersecretary for Research, Education, and Economics, USDA*
11:45 a.m.		Objectives of Breakout Session #1
12:00 p.m.	Dining Room	Lunch
12:45 p.m.	*Various Rooms*	**Breakout Session #1:** Identifying the Biggest Problems in Food and Agriculture
3:00 p.m.	Atrium	Break
3:30 p.m.	Auditorium	Plenary Report Out
4:45 p.m.		Plenary Discussion
5:45 p.m.	Atrium	Reception

TUESDAY, October 3

8:15 a.m.	Auditorium	Objectives for Breakout Session #2
8:45 a.m.	*Various Rooms*	**Breakout Session #2:** Understanding the Scientific Challenges, Knowledge, and Research Gaps
10:30 a.m.	Atrium	Break
11:00 a.m.	Auditorium	Plenary Report Out
11:50 a.m.		Plenary Discussion
12:30 p.m.	Dining Room	Lunch

1:15 p.m.	Auditorium	Keynote: Enhancing Collaboration to Accelerate Agricultural Advancement *Jack Odle, North Carolina State University*
1:55 p.m.		Objectives for Breakout Session #3
2:00 p.m.	*Various Rooms*	**Breakout Session #3:** Scientific Tools and Capabilities Needed
3:15 p.m.	Atrium	Break
3:45 p.m.	Auditorium	Plenary Report Out
4:20 p.m.		Plenary Discussion—Where Are Overlaps and Synergies?
5:20 p.m.		Objectives for Breakout Session #4
5:30 p.m.		**Adjourn for the day**

WEDNESDAY, October 4

8:30 a.m.	*Various Rooms*	**Breakout Session #4:** Science Breakthroughs for Overcoming the Challenges in the Next 10+ Years: Describing Solutions, Scientific Opportunities, and Future Directions
10:00 a.m.	Atrium/Lawn	Break (coffee in Atrium, food on Lawn)
10:15 a.m.	*Various Rooms*	**Breakout Session #4**, continued
11:45 a.m.	Dining Room	Lunch *(Please clear the dining room by 12:30 p.m. to allow staff to prepare for the afternoon plenary session. You are welcome to enjoy our outdoor space for the remainder of your lunch.)*
1:00 p.m.		Plenary Report Out
1:45 p.m.		Plenary Discussion: Issues Raised During Final Report Out

2:15 p.m.	Plenary Discussion: What Have We Missed?
3:15 p.m.	Closing Remarks, Committee Co-Chairs
3:30 p.m.	Adjourn

JAMBOREE WORKSHOP PARTICIPANTS

L. Garry Adams, *Texas A&M University*
Robert (Bob) Allen, *IBM Almaden Research Center*
Ray Asebedo, *Kansas State University*
Vanessa Bailey, *Pacific Northwest National Laboratory*
Julia Bailey-Serres, *University of California, Riverside*
Lance Baumgard, *University of Arizona*
Steve Briggs, *University of California, San Diego*
Jean Buzby, *USDA-ERS*
Don Cooper, *Mobile Assay Inc.*
Mark Cooper, *DuPont Pioneer*
Ryan Cox, *HATponics, Inc.*
Sean Cutler, *University of California, Riverside*
Jorge Delgado, *USDA-ARS*
Daniel Devlin, *Kansas State University*
Michael Doyle, *University of Georgia*
Jillian Fry, *Johns Hopkins University*
Jagger Harvey, *Kansas State University*
Dennis Heldman, *The Ohio State University*
David Hennessy, *Michigan State University*
Georg Jander, *Boyce Thompson Institute*
Xingen Lei, *Cornell University*
Carmen Moraru, *Cornell University*
Dawn Nagel, *University of California, Riverside*
Jack Odle, *North Carolina State University*
Joseph Puglisi, *Stanford University School of Medicine*
Chuck Rice, *Kansas State University*
Carly Sakumura, *NASA Jet Propulsion Lab*
Lisa Schulte-Moore, *Iowa State University*
Soroosh Sorooshian, *University of California, Irvine*
Jim Stack, *Kansas State University*
Laura Taylor, *North Carolina University*
Chris Topp, *Donald Danforth Plant Science Center*
Michael Udvardi, *Noble Research Institute*
George Vellidis, *University of Georgia*

Matthew Wallenstein, *Colorado State University*
Cathie Woteki, *former USDA*
Hongwei Xin, *Iowa State University*

WEBINARS

October 27, 2017	**Webinar: Food Science Research Breakthroughs** Gregory Ray Ziegler, *The Pennsylvania State University* Devin Peterson, *The Ohio State University* John Hayes, *The Pennsylvania State University*
October 30, 2017	**Webinar: Phosphorus Availability and Management** Philippe Hinsinger, *French National Institute for Agricultural Research (INRA) UMR Eco&Sols*
November 3, 2017	**Webinar: Water Resources and Agriculture** Upmanu Lall, *Columbia University*
November 8, 2017	**Webinar: Sensors in Food and Agriculture** Abraham Duncan Stroock, *Cornell University* Suresh Neethirajan, *University of Guelph*
November 9, 2017	**Webinar: Integrating Agriculture into the Built Environment Part I** Peter Groffman, *Cary Institute of Ecosystem Studies*
November 10, 2017	**Webinar: Integrating Agriculture into the Built Environment Part II** Michael Hamm, *Michigan State University* Ed Harwood, *Aerofarms*

Appendix C

IdeaBuzz Submissions Synopsis and Contributors

Instructions for IdeaBuzz

Tell us your idea for innovative research that could elevate the science of food and agriculture. In describing your idea, please comment on how the science and engineering approach you describe might:

- Address a major challenge in food and agriculture
- Create a novel opportunity for advances in food and agricultural science
- Help overcome a technological barrier
- Fill a fundamental knowledge gap that currently holds back progress in the fields of food and agriculture

The responses to IdeaBuzz were reviewed by staff and separated into the following categories: greener plants/crops, animal agriculture, food loss/waste, food safety, resilience/sustainable change, and miscellaneous.

Greener plants/crops (soil, water, land use, climate, plant genetics, phytobiome, etc.)

- Invest in plant-based agriculture
- Understanding soil carbon—carbon sequestration in soil
  - Gaps in science and technology, reliable and affordable mechanisms for testing the carbon content of soil
- Smart food: high-quality foods produced near urban population centers using fewer inputs
- Reinventing potato at the diploid level
  - Genetic gains for higher yield
- Financing should be devoted to applied plant breeding
- Research on cropping systems, including economics, markets, and infrastructure
- Developing hybrid wheat
  - Advanced mating designs, genomic predictions, and chemical hybridizing agents (CHAs)
- Inga alley-cropping
  - Resilient, restores depleted soils
- Soil nutrients need a pathway back to the soil
  - Industrial-scale composting
- Future of food lies in regenerating soil through organic and regenerative farming and land-use practices
- Detox to regenerate
  - Mushroom cultures to detox soil and water. Hemp also good detoxer. Plant everywhere and dredge waterways to apply silt
- Diverse growing systems based on artificial intelligence (AI) harvest
  - Move away from monocultures, and more toward diverse, organic growing systems. In research, work needs to be done on which plants grow well together, and how they can interact to benefit soil/insect health and keep pests away without using any inorganic sprays or products
- Agriculture, biodiversity, and health
- Genetic engineering for sustainable management of crop diseases
  - CRISPR, disease-resistant crops
- Bring agriculture and horticulture sciences closer to home
  - Research, development, and implementation of green roofs, vertical farming, cave farming, hydroponics
- Rebuilding soil microbial communities in agroecosystems
- Soil carbon restoration—regenerative agriculture—our only path forward
- Make vegetables cheaper and easier to eat
  - Invest in social and technological sciences

- Regenerating ecosystem services in grazing ecosystems
  - How different management strategies impact causal mechanisms that drive biological function and socioeconomic outcomes at local and landscape scales
- Crop simulation models that take into account plant genome
- Continuous investment on genomics of crop species
- Deregulation of genome-editing technologies
- Insufficient effort in developmental studies in crops
- Revitalize public breeding programs for interested plant researchers
- More plant molecular biologists and biochemists working on crops
- Opportunities in plant-based meat alternatives
- Restoration of previous grassland soils and revitalizing soils from the woods requires a return to high carbon inputs and nutrient cycling based on turnover of organic matter in synchrony with human, plant, and environmental needs
- Sustainable and resilient agroecosystems in a changing world
  - Sufficient understanding of the mechanisms behind individual behavioral change at the farm/field level, or at the systems level in response to the risks and uncertainties posed by a rapidly changing world
- Harness the power of microbes to enhance agricultural sustainability
- Biofilm control needed for crop cultivation and food safety
- Ingenuity and resiliency of Hopi cropping systems
  - Agriculture techniques to optimize the amount of moisture in the soil
- Making more water available to crops in arid areas
  - Dried and milled *Agave americana* plants contain components that absorb rainwater and keep it being available to crops
- Fertilizers are the key to increasing crop yields around the planet
  - Focus on cropping density, data analytics, remote sensing, drones, seed technology, cultivation tools, pest control, and irrigation
- "Division of labor" among more-specialized crops
  - Nitrogen-fixing legumes could supply all our protein, if they did not waste their limited (C_3 pathway) photosynthate making oil or starch. So breed low-protein maize for high starch or oil yields without nitrogen fertilizer, while breeding low-oil soybeans that yield lots of protein using nitrogen from symbiosis
- New tools to protect our forests from lethal invasive pathogens and insects
- Soil security
  - Protocols for measures, development of technology, valuation of soil as natural capital, evaluation of practices

- Accelerating genetic improvements by cycling of gametes in vitro
- Breeding, research, and production of perennial grain crops and polycultures
- Building a 21st-century soil information platform for U.S. and world soils
- Broadening the range of plants/animals that can be effectively engineered
- Cellulosic ethanol production
- Opportunities for nutrient management
  - Controlled delivery and management of nutrients for plant growth and productivity
- Deep tillage to improve soil hydrologic function and resiliency

Animal Ag: "Greener" livestock (genetics, feed, rumen microbiome, animal nutrition, animal health, climate, environment, etc.)

- Integrate agricultural and ecological sciences to understand pathogen spread
  - Understand AMR bacteria transmission across agricultural–wildlife interface
- Incorporate aquaculture into the discussion of food and agriculture
- Food waste into animal feed
  - New food packaging materials that are digestible by animals, fish, and/or insects
- Insects are more efficient animal feed
- End torture: Ten billion factory farm animals are legally mutilated annually in the United States without any form of anesthetic or pain relief
- Irrational overregulation of transgenic technologies
- Opportunities in clean meat
  - Cultured meat
- Harness the power of microbes to enhance agricultural sustainability
- Accelerating genetic improvements by cycling of gametes in vitro
- Food for 2050 and beyond
  - Cultured meat and other food tissues
- Broadening the range of plants/animals that can be effectively engineered
- Microbiome of the rumen: The time is right for a comprehensive study of the microbiome of the rumen in food animals, including the determinants of colonization of the gut after birth, the role of the microbiome in nutrition and gastrointestinal health, and particularly, its relationship to the animal's immune system

Reduce food loss/waste by half (packaging, processing, distribution, consumer acceptance, etc.)

- Ensuring a safe, secure, and abundant food supply
- Expand urban composting
- Bioremediation
- Retail food waste into animal feed
- Paradigm shifts in fast food feeding and other rapid freezing needs
- Advanced meal processing and preparation
    - Aseptic technologies

Improving food safety (human health, diagnostics, irradiation, consumer behavior)

- Improving human health, nutrition, and wellness of the U.S. population
- Gain more public support for GMOs
- Bioremediation
- Integrate agricultural and ecological sciences to understand pathogen spread
    - Understand AMR bacteria transmission across agricultural–wildlife interface
- Evidence-based decisions empower food policies and consumer health
    - The agricultural, medical, and social science communities need to team up to provide factual, science-based food information in a form easily assimilated by policy makers, professional societies, and consumers
- Agriculture, biodiversity, and health
- Environmental impacts of meeting future human nutrition needs
- Universal in vitro or in silico test for the edibility of a novel substance is a fundamental technology gap in the fields of food and agriculture
- Clean food process technology development
    - Minimal processing technologies
- Biofilm control needed for crop cultivation and food safety
- New tools to protect our forests from lethal invasive pathogens and insects
- Method to rid food safety issues associated with chilled soups
    - E.g., rapid volumetric heating methods
- Nanoscale sensors for food characteristic identification

Pathways for Resilience and Sustainable Change: Identifying key leverage points in the system to effect vast changes needed to ensure success of science breakthroughs and enhance the well-being of society (education and workforce, science communication, data and data sciences/techniques/technology, systems modeling tools, new economic opportunities, drivers, demographics, etc.)

- Support transdisciplinary training grants
- Educate the population on the benefits of plant-based diets
- Adapting and mitigating the impacts of climate change on ag systems
- Smart food: high-quality foods produced near urban population centers using fewer inputs
- New ideas and modeling for urban agriculture
    - Infrastructure and services
- Strengthen the ties between breeders, distributors, and community members
    - Strengthen incentives for feedback between all stakeholders of the product path
- Advancing ag research through data sharing and new data analytics
    - Data sharing, on-farm data research
- Global partnerships for global solutions
    - ASABE (American Society of Agricultural and Biological Engineers)
- Research on cropping systems, including economics, markets, and infrastructure
- Evidence-based decisions empower food policies and consumer health
    - The agricultural, medical, and social science communities need to team up to provide factual, science-based food information in a form easily assimilated by policy makers, professional societies, and consumers
- Agriculture global change challenges
    - Habitat fragmentation resulting from land-use change (expanding agricultural, forestry, and urban areas) leads to biodiversity loss
- Bring agriculture and horticulture sciences closer to home
    - Research, development, and implementation of green roofs, vertical farming, cave farming, hydroponics
- Avoid framing traps: Keeping science and technology in appropriate context
    - Technologism, productivism, efficiency-based sustainability, and reductivism

- Pay farmers for multiple ecosystem services, especially smallholders
- Do not forget the consumer: Engage citizens, applied economists, behavioral economists, and communicators in the development of the priorities
- Regenerating ecosystem services in grazing ecosystems
    - How different management strategies impact causal mechanisms that drive biological function and socioeconomic outcomes at local and landscape scales
- Deregulation of genome-editing technologies
- Irrational overregulation of transgenic technologies
- Train more people in food and agricultural research
- Analytical laboratories for developing countries
    - Need for ways to get soil, water, plant, and other types of samples analyzed continues to be a stumbling block that keep agricultural programs, both research and applied, from moving forward
- Sustainable and resilient agroecosystems in a changing world
    - Sufficient understanding of the mechanisms behind individual behavioral change at the farm/field level or at the systems level in response to the risks and uncertainties posed by a rapidly changing world
- A framework for client-oriented agriculture
- A paradigm shift to agroecology: Context and conservation in agriculture
- All new technologies need a commercialization strategy
    - E.g., Feed the Future Partnering for Innovation program
- Obstacles to big data in plant-level decision making for agriculture
    - Data ownership, data validity, data standardization, data bandwidth, data availability, and model practicality
- The risks of multiple breadbasket failures in the 21st century
    - Need for and movement toward improved probabilistic modeling and prediction of multiple breadbasket failure events and their potential consequences for global food systems
- Soil security
    - Protocols for measures, development of technology, valuation of soil as natural capital, evaluation of practices
- Collaborative research by agricultural, nutrition, natural, and social scientists could reduce this knowledge gap and improve performance across the food system

Miscellaneous

- To make things simpler: Produce food more sustainably and reduce population growth
- Protein synthesis
  - Capture atmospheric nitrogen and convert it to synthetic amino acids and then protein. We should do the research to build efficient bioreactors that combine sustainable synthetic protein with carbohydrates from our agricultural fields to produce healthy, tasty, and sustainable food
- Establishing adequate weather stations for developing countries for crop simulation
- Older ideas also work
  - Organic family farming or permaculture

The committee wishes to thank the following individuals for contributing their ideas via IdeaBuzz (*in alphabetical order*):

Warren A., The Ohio State University
Maureen A. Absten, Natural Health & Energy
Peter Stephen Baenziger, University of Nebraska
VM Bala Balasubramaniam, The Ohio State University
David Baltensperger, Texas A&M University
Verel W. Benson, Benson Consulting
Pierluigi Bonello, The Ohio State University
Paul Brown, Purdue University
Neville Bruce
Marilyn Bruno, Aequor, Inc.
Zack Brym, University of Florida
Edward S. Buckler, Cornell University
Davide Ciceri
Keith Coble, Mississippi State University
D. Curci, Petfinder
Susan Davis, Agavesol (Pty) Ltd.
R. F. Denison, University of Minnesota
Reid Detchon, United Nations Foundation
Jorge Dubcovsky, Howard Hughes Medical Institute
William Fisher, Institute of Food Technologists
Alan Franklin, Research Scientist
Janet Franklin, School of Geographical Sciences & Urban Planning
Alan Franzluebbers, U.S. Department of Agriculture's Agriculture Research Service (USDA-ARS)
Fred Gould, North Carolina State University

Julie Guillen
Christina Hamilton, Experiment Station Committee on Organization and Policy
Mike Hands, Inga Foundation
Jean Hohl
Mitch Hunter, The Pennsylvania State University
Shelley Jansky, USDA-ARS and University of Wisconsin–Madison
Adrienne Job
David C. Johnson, New Mexico State University
Keith F. Johnson, Retired Farmer and Author
Sean Patrick Kearney, University of British Columbia
Anita Klein, University of New Hampshire
Jane Kolodinsky, University of Vermont
Michael Kotutwa Johnson, University of Arizona
Nicola Kubzdela, Student
Timothy LaSalle, International Regenerative Agriculture and Climate
Ken Lee, The Ohio State University
Daniel Magraw, Johns Hopkins University
Andrew P. Manale
David H. McNabb
Joann McQuone, American Society of Agricultural and Biological Engineers
Brian Meyer
Rebecca Milczarek, USDA
Douglas Miller
Roman Molas, Usida R&D Poland
William Mulhern, University of Wisconsin–Madison
Seth C. Murray, Texas A&M University
Carrie Nutter
Jack Odle, North Carolina State University
Stephanie Polizzi
Bob Rabatsky, Fintrac, Inc.
Michael Ramirez
Cheryl Reed
Erin Rees Clayton, The Good Food Institute
Ian Scadden, Texas A&M Agrilife Research
Henry Sintim, Washington State University
Wayne Smith, National Association of Plant Breeders
Mark E. Sorrells, Cornell University
Elizabeth Sparth
Brian Spatocco, Advanced Potash Technologies
Russell Stanton, Zero Aggression Project and the Thrive Movement
Ann Stapleton, University of North Carolina Wilmington and CyVerse

Vala Stevenson, East Village Wellness Circle
Elizabeth Stulberg, Alliance of Crop, Soil, and Environmental Science Societies
Theresa Swarny
Kenneth R. Swartzel, North Carolina State University
Richard Teague, Texas A&M AgriLife Research
John A. Thomasson, Texas A&M University
Michael Tlusty, University of Massachusetts Boston
Michael Twiggs
Paul Vincelli, University of Kentucky
Matthew Wallenstein, Colorado State University
Chandler Wiland
Cathy M. Wilson, Idaho Wheat Commission
Michael Wilson
Robyn Wilson, The Ohio State University